ADVOCATING WEAPONS, WAR, AND TERRORISM

RSA·STR

THE RSA SERIES IN TRANSDISCIPLINARY RHETORIC

Edited by
Michael Bernard-Donals *(University of Wisconsin)* and
Leah Ceccarelli *(University of Washington)*

The *RSA Series in Transdisciplinary Rhetoric* is a collaboration with the Rhetoric Society of America to publish innovative and rigorously argued scholarship on the tremendous disciplinary breadth of rhetoric. Books in the series take a variety of approaches, including theoretical, historical, interpretive, critical, or ethnographic, and examine rhetorical action in a way that appeals, first, to scholars in communication studies and English or writing, and, second, to at least one other discipline or subject area.

Other titles in this series:

Nathan Stormer, *Sign of Pathology: U.S. Medical Rhetoric on Abortion, 1800s–1960s*

Mark Longaker, *Rhetorical Style and Bourgeois Virtue: Capitalism and Civil Society in the British Enlightenment*

Robin E. Jensen, *Infertility: A Rhetorical History*

Steven Mailloux, *Rhetoric's Pragmatism: Essays in Rhetorical Hermeneutics*

M. Elizabeth Weiser, *Museum Rhetoric: Building Civic Identity in National Spaces*

Chris Mays, Nathaniel A. Rivers, and Kellie Sharp-Hoskins, eds., *Kenneth Burke + the Posthuman*

Amy Koerber, *From Hysteria to Hormones: A Rhetorical History*

Elizabeth C. Britt, *Reimagining Advocacy: Rhetorical Education in the Legal Clinic*

Ian E. J. Hill

ADVOCATING WEAPONS, WAR, AND TERRORISM

Technological and Rhetorical Paradox

THE PENNSYLVANIA STATE UNIVERSITY PRESS
UNIVERSITY PARK, PENNSYLVANIA

Library of Congress Cataloging-in-Publication Data

Names: Hill, Ian E. J. (Ian Edward Jackson), 1974– author.
Title: Advocating weapons, war, and terrorism : technological and rhetorical paradox / Ian E.J. Hill.
Other titles: RSA series in transdisciplinary rhetoric.
Description: University Park, Pennsylvania : The Pennsylvania State University Press, [2018] | Series: The RSA series in transdisciplinary rhetoric | Includes bibliographical references and index.
Summary: "Examines commonplace conflicting beliefs that technology will either annihilate humanity or preserve humanity from annihilation. Argues that the paradoxical capacities of weapons influence how humanity understands violent conflict"—Provided by publisher.
Identifiers: LCCN 2018018338 | ISBN 9780271081236 (cloth : alk. paper)
Subjects: LCSH: War and society. | Weapons—Social aspects. | Technology—Social aspects. | Rhetoric. | Paradox.
Classification: LCC HM554.H56 2018 | DDC 303.6/6—dc23
LC record available at https://lccn.loc.gov/2018018338

Printed in the United States of America
Published by The Pennsylvania State University Press,
University Park, PA 16802-1003

The Pennsylvania State University Press is a member of the Association of University Presses.

It is the policy of The Pennsylvania State University Press to use acid-free paper. Publications on uncoated stock satisfy the minimum requirements of American National Standard for Information Sciences—Permanence of Paper for Printed Library Material, ANSI Z39.48-1992.

To Adelaide, Winifred, and Katie,

May your lives be unaffected by fear and violence, along with everyone else's. So say we all.

Contents

Acknowledgments

It is difficult to state fully the respect, admiration, and love I have for my wife, Katie, who puts up with my foibles and shenanigans, and who went so far as to immigrate with me. What will our next adventure be? Snowshoeing?

My daughters, Adelaide and Winifred, provide constant excitement, laughter, rambunctiousness, drama, and love.

Many thanks to my parents for their love and support, especially since I've ended up living in such a far-flung city.

This book would not exist if not for the guidance and support of Ned O'Gorman and Judy Segal. I have the utmost gratitude for your commitment, encouragement, and advice. Thank you.

Thanks to Kendra Boileau, Leah Ceccarelli, and Michael Bernard-Donals—my editors at Penn State University Press and of the Rhetoric Society of America's Series in Transdisciplinary Rhetoric. Special thanks to Leah for providing such insightful commentary and close readings of the manuscript.

Thanks to my friends and colleagues in Illinois and Vancouver with whom I have argued, critiqued, complained, caroused, celebrated, travelled, laughed, mourned, and otherwise lived. In rambling conversations with Peter Campbell, I put some of this book's amorphous ideas into succinct statements. Let's talk again soon.

Thanks to my students at the University of British Columbia with whom I shared my thoughts and arguments about rhetoric and technology, and who responded by challenging them.

The companionship of Hank during all phases of this project deserves special commendation. A walk in the Vancouver rain is never a bad idea, right?

Chapter 2 is derived in part from an article, "Preaching Dynamite: August Spies at the Haymarket Trial," by Ian E. J. Hill, published in *Communication and Critical/Cultural Studies* on 21 April 2016, available online: http://www.tandfonline.com/ http://dx.doi.org/10.1080/14791420.2016.1173220, copyright © National Communication Association, reprinted by permission of Taylor & Francis Ltd, www.tandfonline.com on behalf of The National Communication Association. Chapter 3 includes material that is quoted with permission from the Leo Szilard Papers, Special Collections and Archives at the University of California–San Diego. Partial and early versions of the book's chapters were delivered as papers at conferences, including the 2015 Endnotes University of British Columbia Graduate Conference (chapter 3), the 2014 Rhetoric Society of America Conference (chapter 5), and the National Communication Association's 2011 Annual Convention (chapter 2), 2010 Annual Convention (chapter 4), and 2008 Annual Convention (chapters 1 and 3).

Introduction | Technē's Paradox and Weapons Rhetoric

They will be obedient tools of destruction as long as the spirit contemplates destruction; and they will be constructive as soon as the spirit decides for great buildings.
—Ernst Jünger (1932)

But where danger is, grows / The saving power also.
—Friedrich Hölderlin (1803)

Technology will annihilate humanity—*and* technology will provide humanity's only means of preservation from annihilation. This commonplace belief is what I call Technē's Paradox. This technological and rhetorical paradox appears again and again in popular culture, especially in science fiction, like the *Terminator* movie trilogy (1984, 1991, and 2003) and TV shows like *Battlestar Galactica* (2004–9), *Eureka* (2006–12), and the revived *Doctor Who* series (2005–). In these science-fiction plotlines, technologies annihilate humanity and preserve it. They teach humanity that technology serves us until it kills us, and only what kills us can save us.[1] Rhetorically, such narratives exhort humanity to take heed that its technological hubris will lead to disaster, and they exhort humanity to seek technological solutions for the most catastrophic problems. Technē's Paradox, however, extends well beyond popular culture, permeating the rhetoric of people who ponder the advantages and disadvantages of myriad technologies in real-world, life-or-death situations.

Technē's Paradox influences thought about a broad spectrum of technological dilemmas and their rhetorical deliberation. It undergirds the basic premises of risk analysis implemented by public policy-makers when deliberating the use of technologies vis-à-vis transportation safety, disease, and crime. It galvanizes environmental controversies about pollution, damming rivers, methods of agricultural production, and waste management. Fossil fuels and nuclear fission threaten global catastrophe for the sake of energy production, while wind and water turbines, solar arrays, and long-lasting batteries seem to offer the technological power to avoid or lessen such catastrophes. Technē's Paradox animates assessments of capitalist consumer society and proliferates in descriptions of digital culture and futurism. It manifests when the stakes are not so high, and on the best and worst days, it reflects our exaggerated feelings about the most banal machines. On a hectic morning, the world-ending trauma of a broken coffeepot or a flat tire might make one overly conscious of how people depend on an assortment of "background technologies" for basic mental preservation.[2]

The more dangerous the technology, though, the higher the stakes for negotiating the double truth bound together in the seeming inevitability of simultaneous technological destruction and technological preservation. This book focuses on some of the most dangerous kinds of technologies—weapons—and the rhetoric of their development and use. As Lewis Mumford wrote, "Real orgies of destruction, vast collective eruptions of hate, became possible only when civilization provided the technical means of accomplishing them."[3] Weapons facilitate humanity's "orgies of destruction." The nearly ubiquitous technological and rhetorical predicament of Technē's Paradox and the threat of technological annihilation with weapons pervade humanity's precarious relationship with machines and systems, inflecting our beliefs as well as our actions.

Weapons are designed or appropriated with the explicit intent to kill. Yet in addition to their physical force, they possess geopolitical, ethical, and cultural forces. They are therefore the most critical of technologies for understanding Technē's Paradox. The persistent pervasiveness of Technē's Paradox in this context is observable as an ideology, as an anxiety, and as a rhetoric. As an ideology, Technē's Paradox perpetuates the innovation, production, and deployment of weapons for the purposes of securing humanity from famine, plague, and war. As an anxiety, it presents weapons to the world's population with the permanent threat of catastrophe, even extermination, while it also presents weapons as an anxiety-relieving panacea. As a rhetoric, it constrains and opens up the inventive scope of how people communicate about weapons and how weapons are persuasive. In this book, I argue that Technē's Paradox is fundamental, if not inherent, in the ways that people use technologies to persuade and how technologies themselves are persuasive.

I use the Greek term *technē* to describe this central technological and rhetorical paradox in order to draw attention to the close affinity between the world-making capacities of both rhetoric and technology. In short, technē means skillful craft. Aristotle, by defining rhetoric as a type of technē that involves the discovery and crafting of the available means of persuasion, linked rhetoric with technical activities as diverse as the crafting of health with medicine, the crafting of justice with law, the crafting of boats with shipbuilding, and the crafting of buildings with architecture.[4] Martha Nussbaum defined technē as "a deliberate application of human intelligence to some part of the world, yielding some control over *tuchē* [luck]; it is concerned with the management of need and with prediction and control concerning future contingencies."[5] Universality, teachability, precision, and the "concern with explanation" further characterize technē for Nussbaum.[6] A person possessing technē thereby possesses foresight, "some sort of systematic grasp," and "some way of ordering the subject matter," all of which instill preparedness for and control over technical matters.[7] In these ways, the art of technology and the art of rhetoric overlap, just as the etymological roots of the word "technology" point toward an understanding of this word that welds skill in invention (*technē*) with language (*logos*). The overlap of technology and rhetoric in the idea of *technē* indicates that technological invention influences rhetorical invention, and rhetorical invention influences technological invention.[8] Rhetoric can be considered "the *logos* of *technē*" and the *technē* of *logos*, or the persuasive capacity of both machines and words.[9] Technē, as I use the term, thus refers to the co-creative world-making capacities possessed by rhetoric and by technology, by rhetors and by engineers, and by documents and by machines.[10]

Technologies do not kill alone. And neither do people.[11] Both technology and people depend upon each other as well as upon language, language that is violent and language that motivates violence. Weapons rhetoric, or the ways by which people argue, debate, describe, advocate, and dissent against weapons—as well as how weapons themselves motivate belief and action—is such a language.[12] Weapons rhetoric persuades people to kill through the ideological construction of political economies, religious dogmatism, patriotism, nationalism, racism, and hatred; through military strategists, commanders, and drill sergeants; through politicians, policy, technocracy, and diplomacy; through the mass media; through burgeoning weapons industries and capitalist expansion; through the militarization of science, engineering, and universities; through the marketing and selling of weapons to individuals, groups, and states; and through the physical presence of weapons.

There is a growing body of scholarship on weapons rhetoric, including Lisa Keränen's work on how "bio(in)security" promotes technological solutions for biological weapons attacks; Gordon Mitchell's work on how "strategic deception" navigates and suppresses public debate about weapons systems; and Edward Schiappa's work on how "nukespeak" discourages public involvement in nuclear policy-making.[13] Moreover, Paul Virilio has been adamant that violence is inherent in communication, and that the "information bomb," or "weapon of mass communication," is not mere wordplay.[14] At each moment when weapons have been deployed, and at each deliberative moment when peace has degenerated into war and when war has stuttered to a halt, people have wondered about whether weapons would cause humanity's annihilation or its preservation. Weapons rhetoric is language that perpetually comes into conceptual contact with Technē's Paradox. Thus, to speak or write about weapons has prompted thinking about how rhetoric helps spur people toward perpetual global peace or extermination.[15] In turn, I call the various ways rhetors adapt their language to Technē's Paradox "negotiating the Paradox." As rhetors negotiate both the presence of weapons and the enigma of Technē's Paradox, the fate of larger and larger populations depends on the power of rhetoric.

1. The Dilemma of Technē's Paradox

The simultaneously preservative and destructive dual uses of nitrogen fertilizers are an initial example of how weapons, and technologies in general, summon Technē's Paradox. At about nine o'clock in the morning of April 19, 1995, a truck bomb ripped into the Alfred P. Murrah Federal Building in Oklahoma City. The explosion traveled at 7,000 miles per hour and its force of 74,000 pounds equaled the force of a magnitude 4 earthquake.[16] The primary explosive, ammonium nitrate fertilizer, demolished the building's façade. The bombers, Timothy McVeigh and Terry Nichols, constructed the bomb by loading 108 fifty-pound bags of the fertilizer, three five-hundred-pound drums of nitromethane fuel, and some ammonium nitrate/diesel fuel oil (ANFO) explosives into a rental truck.[17] They killed 168 people and injured 680. The surprise attack, which came in response to the US government's 1993 raid of the Branch Davidian compound outside Waco, Texas, became, in an instant, a shocking reminder of synthetic fertilizer's explosive power. The bombers did not convert fertilizer into a weapon so much as they implemented one of ammonium nitrate's primary uses—killing lots of people.

German chemist Fritz Haber, who in 1918 won the Nobel Prize in Chemistry, began researching how to make nitrogen fertilizers in the early twentieth century with the goal of feeding more people, not killing them. In 1911, the technology to mass-produce ammonium nitrate must have seemed to Haber like an unequivocal boon for humanity. Haber said of his discovery that "improved nitrogen fertilization of the soil brings new nutritive riches to mankind" to the extent that "the chemical industry comes to the aid of the farmer who, in the good earth, changes stones into bread."[18] However, once World War I broke out, Haber applied his ingenuity not only to feeding a hungry wartime population but also to exploding the enemy. By inventing an industrial process that combined nitrogen and hydrogen to create ammonia, he helped accomplish both feats. The Haber process, later refined by Carl Bosch, supplied Germany with a significant portion of the ammonium nitrate it needed to arm itself for four years of intensive warfare. Furthermore, Haber's chemical acumen prepared him to oversee the first successful German chemical-warfare attack, and the first successful chlorine-gas attack, at Ypres, on April 22, 1915. In 1916, he became the director of Germany's Chemical Warfare Service, and his *Leuna-Werke*—a vast chemical factory complex—provided Germany with chemical munitions and fertilizer as the primary producer of each.[19] Thus, the same chemist whose technology empowered a massive increase in agricultural production in the twentieth century also empowered the widespread development of military explosives, the German push to inaugurate chemical warfare, and the ensuing international arms race. The huge expenditures on German chemical plants to produce dual-use nitrates propelled the expansion of the military-agricultural technology, an expansion that postwar population growth and agricultural necessity would make permanent, even with the destruction of some German chemical plants as per the dictates of the Treaty of Versailles.[20] At the end of the twentieth century, ammonium nitrate both nourished the monoculture of Oklahoma's cash crop—wheat—and exploded Oklahoma City's Murrah Building. It nurtured, and then it destroyed. It continues to do both.

Yet the ingenuity of the fertilizer truck bomb itself and its appropriation of commonplace industrial products did not result in the weapon's vilification, in contrast to, for example, the vilification of hydrogen bombs and sarin nerve gas. To malign the mass production of ammonium nitrate, nitromethane, and rental trucks would be to disparage three fundamental components of the US economy—agriculture, chemistry, and transportation. To bar the products of German war technology from common use would threaten large populations with starvation.[21] So should people live in fear of fertilizer or use it as definitive proof that humanity can invent technology to sustain itself? Not just fertilizer: all technologies confront people with this dilemma, which requires assessments of their advantages and disadvantages.

Technē's Paradox has influenced humanity's concept of technology since at least ancient Greece and has a complex rhetorical ancestry. Protagoras's cosmogony, as reproduced by Plato, demonstrates that technology has long made humanity think about its preservation and destruction at the same moment by tying human survival to technological achievement. Protagoras depicted both how weapons fail to ensure survival and how technologies, in the form of political organization and law, can ensure survival. The sophist recounted that people "were devoured by wild beasts, since they were in every respect the weaker, and their technical skill, though a sufficient aid to their nurture, did not extend to making war on the beasts, for they had not the art of politics, of which the art of war is a part."[22] Even when gathered together for protection, early humans would kill each other, which made Zeus fear "the total destruction of our race."[23] The fate of humanity seemed to rest on its technological capacity either to preserve itself or to cause its own demise by failing to withstand war and nature. Ancient technologies neither guaranteed survival nor produced self-annihilation, but the concept that the innovation of technologies could empower one population to exterminate another or to preserve itself in the face of annihilation became a norm of technology discourse. This scheme of ideas began to inform how rhetors expressed technology and, in turn, the proliferation of Technē's Paradox in weapons rhetoric has reiterated and reinscribed the concept such that it has acquired an aura of permanence, immutability, and inescapability.

This ancient dilemma filtered into the twentieth century's "technological society," according to Jacques Ellul. His essay "The Technological Order" proposed four technological rules that reiterate Technē's Paradox:

1. All technical progress exacts a price;
2. Technique raises more problems than it solves;
3. Pernicious effects are inseparable from favorable effects; and
4. Every technique implies unforeseeable effects.[24]

Ellul further asserted that "it cannot be maintained that technical progress is in itself either good or bad. In the evolution of Technique, contradictory elements are indissolubly connected."[25] As much as any machine exists within the vast system of technology that is enmeshed with human activity, no one ought to claim that it is either a pure boon or detriment to everyone, although these are both common assessments. In the words of Harry Collins and Trevor Pinch, technology has a tendency to seem "all good or all bad."[26] What is humanity's technological fate, asked F. Buckminster Fuller, "utopia or oblivion?"[27] According to Technē's Paradox, humanity has decided to imagine the answer as both.

In another recent reiteration of Technē's Paradox that resembles the pervasiveness of Ellul's four rules, Ulrich Beck's "risk society" manages the uncertainty that accompanies new technologies.[28] The Paradox takes a central role in any rigorous attempt to address how people use technologies when those technologies are conceived of as fraught with both benefits and dangers. According to Beck, "modern" humans face a high probability that they will exterminate their own species with chemical, nuclear, genetic, and ecological technologies.[29] Although industrialization spurred the development of beneficial technologies that, in a type of circular reasoning, then legitimated their invention, those once-beneficial technological projects now confront humanity with disaster. Beck therefore contended that the risk society must use "risk logic" to distinguish "between risks and threats" in order to organize politics and formulate policy.[30] Risk analysis, which delineates the positive and negative trade-offs caused by technologies, reinscribes the tension between preservation and destruction for both local communities and the entire earth, depending on the risk and threat level ascertained during technological policy debates. Beck argued that, with catastrophe looming over the global population because of the panoply of risky technological decisions, savvy technological political design might stave off disaster if the dilemma gets solved. But it never does.

Thus, Technē's Paradox refers not just to a dilemma, but also to the manner of thought that it developed and a persistent rhetoric—all of which have a continuing history. It has become a foundational figurative and argumentative principle of technology. Philosophies, sociologies, and histories of technology, in addition to interpretations of modernity, culture, society, civilization, industry, economics, and politics, all must address Technē's Paradox, as must the politicians, engineers, soldiers, activists, scholars, and others who, concerned with the practical consequences of technologies, write and speak about weapons. In this sense, rhetors can negotiate Technē's Paradox by reiterating it as a general principle, reimagining and using it for specific contexts, rejecting it, ignoring it, or tweaking it as they wish. All of these ways of negotiating Technē's Paradox proliferate it as a flexible but persistent concept.

In the remainder of the book, I argue that over the past two centuries, weapons advocates have negotiated Technē's Paradox by delving into the argumentative and rhetorical instability of paradoxical thinking, by attempting to stabilize instability, and by attempting to circumvent it. By rhetorical instability, I mean the ways that rhetoric helps to keep meaning, knowledge, truth, and ethics in flux even more so than in the often inconsequential, everyday ways that language "can be moulded into one shape after another."[31] Consider how Evelyn Fox Keller framed the instability of language. According to Keller, "the dynamic interplay among the different meanings a term may connote" indicates how language is characterized by "variety," "plasticity," "imprecision and flexibility," and "slippage."[32] The "versatility" of every word can be "malleable" and "unstable" enough to support a "multiplicity of uses" as well as opposed arguments.[33] Keller did not argue that a word's meaning cannot ever be stabilized in a particular context or have a literal interpretation, but the capacity for meaning to be temporarily stabilized by multiple redefinitions in different specific contexts causes further instability.[34] In short, as Michel Foucault wrote, words "are burdened with meanings that they cannot control."[35] Paradoxical weapons rhetoric is such a realm of difficult-to-control rhetorical instability.

2. A Rhetorical Approach to the Paradox of Weapons Rhetoric

To show the ongoing vibrancy of Technē's Paradox in weapons rhetoric, I implement a type of rhetorical analysis that John Angus Campbell dubbed a "longitudinal case study." According to Campbell, this method studies the "kairotic occurrences" that "become self-constitutive chapters in the anthology of meaning we call history."[36] Campbell thought that juxtaposing and combining Michael Leff's close textual analysis with Michael Calvin McGee's ideographic analysis provides a way to connect analyses of chronologically disparate texts to create an overarching historical claim.[37] A synthesis of both methods—of close textual analysis, which unfolds "in a specific situation and discloses its meaning within textual time," and of ideographic analysis, which unfolds and gathers "meaning over centuries"—creates a "House of the Middle Way."[38] Whereas McGee's method considers the "rhetorical artifact as but a partial expression of a larger cultural whole" that demonstrates how "invention is grounded in history and history is a matrix of possibilities," Leff's method focuses on "the local stability of the text" in question.[39] Campbell concluded that the "house of the middle way" shows that "meanings . . . tend to stabilize along specific rhetorical-cultural axes." In turn, "we come to understand the rhetorical meaning in larger historical-interpretive patterns."[40]

This book's methodology is similar to Campbell's longitudinal case study. My close textual analyses of specific documents show how the current relationship between humanity and weapons was rhetorically and technologically invented in a historical period that began with the confluence of new, large-scale warfare and the publication of Thomas R. Malthus's population theory in the late eighteenth and early nineteenth centuries, and continues through the end of the twentieth century with the terrorist bombing campaign of technophobe Ted Kaczynski, also known as the Unabomber. These instances of weapons interacting with rhetoric, and three additional cases—accused Haymarket-bombing conspirator August Spies's preaching of dynamite (1886), the humane, preservative power of chemical weapons advocated in Maj. Gen. Amos A. Fries and Clarence J. West's US Army textbook *Chemical Warfare* (1921), and the rhetorical being of Manhattan Project physicist Leo Szilard—all took up popular attitudes and arguments about dangerous technologies, reinvented them, and radiated the updated versions outward to influence further innovation. I therefore concur with Campbell's assertion that in a longitudinal case study, "each construction absorbs the meanings of its context and bodies those meanings forth in a local explosion of rhetorical energy."[41] Such local explosions indicate both the enduring influence of Technē's Paradox and the sometimes stable and sometimes changing historical trajectory of weapons rhetoric. By following Campbell's methodological approach, I can account for how specific weapons and weapons advocates created rhetorical situations as they moved across time, and how weapons rhetoric radiated into the present from "key discursive epicenters."[42]

A second methodological aim of this book is to extend Campbell's concept of the middle way in order to occupy a transdisciplinary space that conjoins rhetoric and science and technology studies (STS) in a way that bolsters the material aspects of technology criticism while still keeping them tethered to language.[43] To bring rhetoric and STS together with a critical approach that meshes with Campbell's "house of the middle way" rhetorical methodology, I draw upon the two disciplines' shared and growing respect for materialist critiques that rethink how things interact with people and how technologies, as objects, exert force on other objects or are "political agents" with "thing power" that impinge on human affairs.[44] By treading this middle-way path, I aim to bring together insights from textual, material, and hybrid theories without attempting to offer some sort of ideal "golden mean" between language and objects.[45]

My middle-way methodological approach to materialism and textuality is to analyze what being in the presence of weapons entails in relation to what is entailed by being in the presence of words about weapons. With this middle-way approach, I aim to demonstrate that greater attention to rhetoric could broaden the extradisciplinary appeal of STS, and that greater attention to STS could broaden the extradisciplinary appeal of rhetoric.[46] "Mixtures of humans and nonhumans" or "actor-networks" that consider the nonhuman agency of materiality in conjunction with human communication provide, as Bruno Latour has written, "*other resources* than the ones classically handed down . . . by rhetoric."[47] In such networks, technology and rhetoric are both "liminal," so this book seeks a "reflexive" approach that acknowledges the influence of both technology on language and language on technology.[48] Classical rhetoric and text-based rhetorical criticism thus can meet machines and mixtures of humans and nonhumans on the path of the middle way. Writing about "eloquent objects"[49] without reference to their meaning is a challenge.[50] My middle-way path does not reject textual criticism and is therefore not fully identifiable with "the speculative turn," which seeks to examine objects without resorting to human meaning as a means of explanation.[51] My methodology is rooted in the material and the textual, but there is no singular middle-way path by which to understand the co-creative aspects of rhetoric and technology. Each chapter therefore takes a somewhat different approach to the middle way to indicate different means by which STS and rhetoric can converge via a materialist perspective toward eloquent objects.

Paying attention to the presence of weapons and the presence of words about weapons unites the material with the textual, and scholarship from both STS and rhetoric has called for such transdisciplinary projects. From the perspective of STS, Karen Barad and Lorraine Daston bring rhetoric and STS into contact via new-materialist perspectives.[52] According to Barad, "matter and meaning are not separate elements," such that even "the smallest parts of matter are found to be capable of exploding deeply entrenched ideas and large cities."[53] Nuclear physics—one such manifestation of meaning and matter—altered humanity's understanding of nature and blew up Nagasaki and Hiroshima. Barad therefore brings together objects and rhetoric to allow "matter its due as an active participant in the world's becoming, in its ongoing intra-activity," while also providing "an understanding of how discursive practices matter."[54] Lorraine Daston concurred that "things knit together matter and meaning."[55] For Daston, when things speak, they possess a rhetorical capacity. "Even if they do not literally whisper and shout, these things press their messages on attentive auditors—many messages, delicately adjusted to context, revelatory, and right on target," she wrote.[56] Thingly speech is audible to humans according to each entity's materiality, and thereby possesses a type of rhetorical capacity to motivate people. "Like seeds around which an elaborate crystal can suddenly congeal, things in a supersaturated cultural solution can crystallize ways of thinking, feeling, and acting," Daston explained.[57] Being in the presence of a thing is to be in the presence of rhetoric, whether or not anybody supplies words about the thing.

The theoretical impetus behind my methodology is not to purge either the human or the nonhuman from criticism, but to look at their copresence, their coinventive capacities, or eloquent objects' "alliances with humans."[58] After all, "the realist turn" can temper humanistic criticism's centripetal focus on language by rethinking the interactivity of human and nonhuman agencies.[59] Weapons influence humans and humans use weapons to influence. To get at an understanding of eloquent objects, one need not abandon either the insights granted by language to study objects or the insights granted by objects to study language.[60] As Campbell suggested, the middle way should incorporate the textual and the material. Thus, in between "the rhetorical turn" and "the material (or speculative) turn" is the middle way, a place inhabited by "rhetorical materials," "material-rhetorical assemblages," "rhetorical ecologies," "orator-machine assemblages," the mangled "dance of agency" that takes place between human and machinic "material performativity," and human and nonhuman "actants"—rhetors and weapons—that motivate belief and action.[61] By examining the interaction of words and weapons, I examine their alliance, their copresence, and their coinventiveness. When weapons talk, they do rhetoric. Therefore, as the book unfolds, each chapter includes a middle-way section that "follows, traces, and describes" weapons as material forces that impinged upon weapons advocates as part of the "larger background's" nonhuman "assemblage of elements."[62]

Granting technology rhetorical power and granting rhetoric technological power both point toward a technological conception of rhetoric that diverges from the definition of rhetoric as a verbal activity performed only by humans. If the rhetoric of technology consists of the ways that people argue, debate, advocate, dissent, and otherwise describe machines and systems, then a technological rhetoric, or machine rhetoric, must combine the rhetoric of technology with the idea that technologies have the capacity to induce human beliefs and actions independently of any human intention on the part of designers, engineers, users, and rhetors. If, as McGee suggested, each human society formulates a distinctive idea and practice of rhetoric to suit differing material and cultural contexts, then, as Richard McKeon suggested, the contemporary technological age requires a technological reformulation of rhetoric.[63] Rhetoric in the "technological age" should therefore use the conceptual and terminological "remnants" from the history of rhetoric that remain pertinent, while diminishing or suppressing unhelpful philosophical baggage from earlier periods, and it should recognize contexts in which differing conceptions of rhetoric do or do not function.[64] For most of its history the discipline of rhetorical studies has emphasized the agency of humans. Hence, the discipline has been cautious about taking up the idea that nonhumans either share agency with humans or possess it themselves. But either applying the idea of nonhuman agency to the act of rhetoric or dismissing the importance of agency is just the sort of tinkering with traditional rhetorical concepts that machine rhetoricians should do.[65] As eloquent objects, weapons have much to say to us. Timothy Morton's definition of rhetoric provides a useful way to think about how we listen to technologies. He defined rhetoric as "what happens when there is an encounter between any object, that is, between alien beings."[66] When they have "momentum," these entities make it difficult to determine whether humans or technologies are more in control.[67] So in the analysis that follows, if weapons appear to possess a type of agency to motivate belief and action, then such a "careful course of anthropomorphization," in the words of Jane Bennett, indicates when technologies possess rhetorical force.[68] But when weapons are reticent and seem to withdraw from us,[69] words about weapons provide another type of rhetorical presence. Thus, conceiving the middle way as the interactive conjunction of the material and the textual can build upon the promise of conceiving rhetoric as something that both people and technologies do.[70]

The primary way I tinker with the concepts of rhetoric, technology, and materiality is by emphasizing the importance of rhetorical presence. Presence was a key term for Chaim Perelman and Lucie Olbrechts-Tyteca, who used the term to refer to the use of words to bring a distant object or idea closer to an audience, the use of words to enhance the importance of a present object, and the persuasiveness of objects. "*Presence* acts directly on our sensibility," they wrote, sounding a bit like British philosopher John Locke's theory of communication.[71] Rhetoric scholarship has tended to emphasize the verbal aspect of presence, but Perelman and Olbrechts-Tyteca also noted that "concrete objects" create or enhance presence.[72] They described orators who presented objects to audiences to enhance their verbal persuasive force—such as Antony waving Caesar's bloody tunic before the people of Rome—to exemplify concrete, objective presence in action.[73] Divorcing this second, objective type of presence from its verbal and oratorical associations is a useful way to begin to understand how weapons and other objects possess a motivational force unto themselves. With or without the presence of words about weapons, weapons possess their own rhetorical presence. Virilio implied the importance of the rhetorical presence of weapons for those who undertake analyses of violent conflict. "Weapons are tools not just of destruction but also of perception—that is to say, stimulants that make themselves felt through chemical, neurological processes in the sense organs and the central nervous system, affecting human reactions and even the perceptual identification and differentiation of objects," he wrote.[74] That weapons—and technologies in general—have rhetorical presence is perhaps the most fundamental aspect of machine rhetoric that aligns the rhetorical study of technologies with STS scholarship. Therefore, to approach the middle way where both humans and technologies motivate action and belief, each chapter examines both what is entailed by being in the presence of weapons with a materialist analysis and what is entailed by being in the presence of words about weapons with a textual analysis.

What vexes the materialist analysis of the interaction of Technē's Paradox and weapons rhetoric are the very human volitions of Malthus, Spies, Fries, Szilard, and Kaczynski. When weapons motivated them to act, they developed historical, political, and technological aspirations that crashed their activities into Technē's Paradox. These figures not only desired to make important arguments about weapons for specific political ends, but also wanted to entrench their arguments into how society organizes itself. By mobilizing their understandings of how to think about and use weapons, these rhetors sought to resolve controversies surrounding the presence of weapons and to create solutions for some of their respective societies' most pressing problems.[75] The weapons rhetors who inhabit this book attempted to generate influential weapons epistemologies based on their political stances and negotiation of Technē's Paradox, and they aimed to alter society to resemble their attitudes toward weapons. Whereas Malthus wanted to influence the development of British economics and governmental class organization to reform and perpetuate early industrial capitalism, Spies desired to martyr himself to spur the extinction of capitalism. And whereas Fries and West fought to justify the permanent place of chemicals in wars, household consumer products, and the economy, Szilard sought to use the Bomb to construct a world of peace. Kaczynski desired to destroy all aspects of technology in society and thereby to rupture, apocalyptically, the course of human events. Thus, all of these actors imagined themselves either implicitly or explicitly as historically important, and their importance hinged upon their success in using and advocating weapons. In turn, their relative successes and failures indicate how weapons constrained and freed their individual capacities to change society by limiting the scope of their activities or suggesting new ways to use weapons to motivate people.

3. The Longitudinal Development of Weapons Rhetoric

In order to demonstrate the dispersed replication of paradoxical weapons rhetoric, the following chapters recount how weapons rhetors both grasped backward at rhetorical tactics and strategies to reinvent or redeploy them, and invented tactics and strategies better calibrated to the presence of new weapons. Hence, these cases show synchronic rhetorical accounts of specific moments in the history of weapons that, when combined, show a diachronic account of weapons rhetoric across time.

In chapter 1, I examine Thomas Malthus's *Essay on the Principle of Population* to further describe and define Technē's Paradox. I argue that Malthus's population theory exemplifies the process of rhetorical invention that centralized the melding of faith in technology as humanity's preservation with anxiety about potential technology-induced annihilation. Multiple historical events, technological innovations, and political movements at the turn of the eighteenth century make Malthus's population treatise an apropos pre-text by which to gauge and explain the importance of Technē's Paradox for subsequent discourse about weapons technology. By pre-text, I mean both a conceptual frame with which to examine such texts and a kind of living rhetorical frame that has animated weapons advocacy. Malthus lived as the so-called agricultural and industrial revolutions increased humanity's productive capacity to unheard-of levels, which led him to wonder what would happen to large populations if they descended into a global war caused by the cyclical success and failure of agricultural technologies. For this chapter's middle-way argument, I suggest that the embodied rhetorical presence of the population bomb—exemplified by population growth and the enormous standing armies of the Napoleonic wars—entailed that large portions of the global population could be deemed "redundant."[76] As I analyze his *Essay on the Principle of Population* as a paradoxical pre-text, I demonstrate that Malthus used three rhetorical tactics important for the longitudinal history of weapons rhetoric. First, his use of questionable agricultural and demographic statistics helped to establish his "principle of population" as a natural law by presenting sublime numerical ratios that depicted improbable environmental scenarios well beyond the potential of sustainability. Second, when Malthus speculated about the catastrophic results of overpopulation, his generalizations abstracted and amplified the magnitude of the results to threaten the entire earth. Third, metaphorical *antistasis*, or the "repetition of a word in a different sense" so that each usage alludes to different connotations of the term, empowered Malthus to drift between the preservative and destructive tendencies of machines without contradicting himself.[77] Not only did these rhetorical tactics undergird the plausibility of his "principle of population," but they also indicated how weapons rhetors would take up concerns about humanity's technological fate. Thus, as a pre-text, his work provides an entry point for understanding how rhetoric impinges upon weapons and how weapons impinge upon rhetoric.

Chapter 2 examines how August Spies preached dynamite. Spies, the accused anarchist archconspirator of the infamous Chicago Haymarket bombing, negotiated Technē's Paradox by delving into and exploiting the rhetorical instability of dynamite. In 1886, dynamite seemed to suffuse the entirety of Chicago, and my analysis focuses on the crucial object of the Haymarket events—dynamite—as much as it focuses on Spies's invocations of the weapon. Being in the presence of dynamite entailed that both Spies and dynamite had partial control over how the Haymarket trial unfolded, I suggest in this chapter's middle-way argument. Dynamite's ubiquity and the supposed threat to humanity posed by one explosive device guided public anxiety about terrorism. And in his final courtroom address, Spies provoked that anxiety as he espoused anarchism, advocated political violence, and argued that dynamite would preserve humanity from capitalism. Two primary rhetorical tactics characterized Spies's preaching of dynamite. First, Spies strategically used the polysemous meanings of the term "dynamite" to distance his calls to arms from the Haymarket bombing. Second, warranted by the presence of dynamite, he made elaborate turnaround arguments based on the preservation and destruction of capitalism and the preservation and destruction of workers. Spies thus used the rhetorical instability of dynamite to provoke political instability, a strategy that his colleagues would denounce upon Spies's execution.

In chapter 3, I turn from dynamite bombs to the proliferation of chemical weapons during World War I and after. In order to advocate mustard gas and other chemical weapons, Maj. Gen. Amos A. Fries and civilian scientist Clarence J. West did not exploit the instability of weapons rhetoric. Rather, I argue, Fries and West attempted to stabilize chemical-warfare discourse and render Technē's Paradox inert by categorically declaring chemical weapons to be preservative. As I show in this chapter's middle-way analysis, Fries and West faced widespread, vitriolic condemnation of chemical warfare based upon the recalcitrant material presence of gas as that presence manifested in the physiology and psychology of gas-attack victims. In response, Fries mounted a public-relations campaign that attempted to reshape public opinion about chemical weapons by framing them as the humane and efficient way to attain victory. As part of this campaign, Fries and West's *Chemical Warfare* utilized two prevalent rhetorical tactics to stabilize the Paradox. First, in order to make mustard gas sound more appealing and to ease rampant fear of the new weapons, they used statistical proofs and comparisons to conventional weapons in order to downplay chemical warfare's physiological and psychological effects. Second, amid the unprecedented destruction of World War I, they amplified the power of gas with totalizing generalizations that elucidated the enormity of the new weapons' potential preservative efficacy on a global scale.

In chapter 4, I examine the writings and activities of atomic-bomb physicist Leo Szilard to argue that he negotiated Technē's Paradox simply by being paradoxical. An important figure in the Manhattan Project, he reversed his pro-Bomb attitude and began lobbying for its abolishment and arms reduction once he realized that the engineers and scientists working at Los Alamos would soon construct the Bomb. In the final years of World War II, recognizing the dire political consequences of nuclear fission, Szilard attempted to situate nuclear physicists—especially himself—between the idealistically truthful realm of science and the practically manipulative realm of politics in order to gain political leverage. Szilard asserted that because atomic physicists had devised the means of massive destruction, they alone were most capable of devising how to use the Bomb to bring peace. The paradoxical rhetorical being he created for himself embodied a manipulated version of the normative scientific ethos to carve out a place for scientists in politics, a collective scientific voice to give credence to atomic physicists' policy proposals, and a realpolitik attitude to propel Bomb research and policy. In this chapter's middle-way argument, I suggest that the material presence of the Bomb authorized Szilard to rearrange atomic facts, both scientific and fictional, to suit changing political, scientific, and imagined exigencies.

Chapter 5 examines the bombs and writings of Ted Kaczynski, especially his treatise *Industrial Society and Its Future*, which is better known as *The Unabomber Manifesto*. Like Fries and West, Kaczynski negotiated Technē's Paradox by attempting to stabilize the instability of weapons rhetoric. But in contrast to Fries and West's attempt to do so by categorically defending a weapon as a preservative force for humanity, Kaczynski categorically rejected the position that any technology can be preservative. Instead, he sought stability by insisting the opposite position—that technology is categorically a force of human annihilation. Therefore, he recommended revolution against technology. In tandem with his improvised explosive devices (IEDs), Kaczynski used several rhetorical tactics to stabilize Technē's Paradox. First, he diminished all nontechnological problems as distractions from his central premise that humanity must revolt against technology. Second, he used the figure of diminishment to refute and downplay the beneficial aspects of technology, and third, he amplified the destructive aspects of technology. Kaczynski's rhetoric displayed an unwavering, inflexible, and unyielding quality. His rhetoric was hard, by which I mean that Kaczynski did not waver from his certain and ruthless attack on machines, systems, and technologists. This type of hard machine rhetoric forced Kaczynski's audience to make a categorical and irreparable choice either to acquiesce to a future of technological disaster or to revolt against it. I argue in this chapter's middle-way section that upon his arrest, Kaczynski's IEDs supplanted his agency. The presence of his eloquent IEDs dictated the government's reaction to his rhetorical violence, entangled numerous people in his agenda, spoke terror rather than solidarity, and bifurcated his persona into that of a reasonable technology skeptic and a mad bomber.

In the book's conclusion, I ruminate on the general applicability of Technē's Paradox to all technologies. I then show how the examination of these influential texts, rhetorical tactics, and weapons reveals some overriding rhetorical strategies. When taken together, the rhetoric and longitudinal history of these case studies shows that the commonplaces of late twentieth-century and early twenty-first-century weapons rhetoric have origins much earlier than either the advent of the Bomb or the War on Terror. Although specific rhetorical tactics and strategies may fade from use with the innovation of certain weapons, all of the arguments, figures, and tropes used by Malthus, Spies, Fries and West, Szilard, and Kaczynski are still prevalent. Thus, I aim to add to the already-vexed task of debating weapons, terrorism, and state violence a technological layer that remains vital in the rhetoric of gun control, suicide bombings, and drone strikes.

In order to "brush history against the grain," as recommended by Walter Benjamin, critics must confront the barbaric rhetoric of successful domination as well as the failed prescriptions, reforms, and revolts that have aimed to remedy humanity's propensity for violence.[78] Hence, I conceive of my critical role as not only aiming to confront the rhetorical and technological invention of weapons in the past, but also to confront speculative prescriptions for change. This study aims yet another level of criticism at the construction and maintenance of weapons, despite their apparent resistance to reformation. The historical trajectory of weapons rhetoric provides an opportunity to judge the successful and failed tactics, strategies, ethics, and ideas that have helped to shape how individuals and states use violence.

With the world's attention now attuned to conflicting and uncertain probabilities that local populations, humanity, and all other life could survive a war fought with so-called ultimate weapons, the negotiation of Technē's Paradox continues to characterize weapons rhetoric. From individual weapons that threaten our communities, like the dynamite bombs of Chicago, Oklahoma City's fertilizer truck bomb, the assault rifles of Sandy Hook Elementary School and the 2017 Las Vegas shootings, and the handguns of the Emanuel African Methodist Episcopal Church shooting, to the more destructive weapons that loom over international conflicts—North Korea's nuclear missiles being just one example—the successes of weapons advocates who design, manufacture, and promote weapons motivate death and destruction. Weapons are fearsome, so how we make them, which weapons we make, how we sell them, how we deploy them, and how we use rhetoric to negotiate not just Technē's Paradox but also the rhetoric of people who make weapons happen—from the inventor of the fuse to the person who lights it—should be judged and people should be held accountable, as should the weapons themselves. Weapons represent much more than their own material existence, and rhetoric helps to freight them with psychological, cultural, and ideological weight. And when weapons and rhetoric push upon the intangible limit between persuasion and violence, those moments of argument, debate, advocacy, and dissent help to define the character of the conflict that ensues—how and when people will live or die.

1

Thomas Malthus's Population Bomb as a Pre-Text for Technē's Paradox

It must be acknowledged that bad theories are very bad things, and the authors of them useless and sometimes pernicious members of society.
—Thomas R. Malthus (1803)

A man like me does not give a shit about the lives of a million men.
—Napoleon Bonaparte (1813)

The Reverend Thomas Robert Malthus hurled this chapter's first epigraph at the many critics of his *Essay on the Principle of Population* (1798).[1] Malthus's counterattack mirrored his critics' accusations that his population theory presented British society with dangerous ideas from a dangerous advocate. According to Malthus, human fertility outpaces the technological, political, and economic capacities to sustain population growth. With this general law as a premise, Malthus aimed to abolish the Poor Laws that had provided relief to paupers since 1597. According to Malthus, the poor were already destined for lives filled with "misery and vice."[2] So Malthus implied that they must be "checked" rather than assisted, lest their procreative power drain state funds, deplete natural resources, foment revolution, or otherwise cause calamity. The *Principle of Population* thus pitted economic and class interests against each other amid the constraints of populations competing for basic biological survival. To support the abolishment of the Poor Laws, Malthus sought to prove the perniciousness of helping the poverty-stricken masses. Controversy erupted. Malthus was labeled the pernicious one.

I argue that Malthus's role in the controversy helped to establish the idea of Technē's Paradox in popular thought via his historically embodied rhetoric, which melded simultaneous depictions of humanity's total destruction and total preservation with statistical amplification, universal generalizations, and the unstable meaning of the word "machine." This melding brought the global population face to face with its two polarized ultimate technological fates and, thereby, Malthus provided some basic commonplaces of paradoxical weapons rhetoric. While drawing attention to the concept of the population bomb, this chapter aims to show how *An Essay on the Principle of Population* functioned as a pre-text of weapons rhetoric. By pre-text, I mean an artifact that serves as an initial manifestation of a specific rhetorical stance and that clarifies a set of argumentative claims, which in turn become a population's ideological and rhetorical commonplaces. The pre-textual significance of Malthus's statistics derives from their portrayal of human death on a massive, global, and sublime scale; the pre-textual significance of Malthus's generalizations derives from their provision of an argumentative template for demonstrating humanity's technological and military imperilment; and the pre-textual significance of Malthus's polysemous use of the word "machine" derives from the way the term incorporated weapons into paradoxical thinking about the preservation and destruction of humanity. In short, Malthus prepared humanity for the idea that weapons exemplify the most extreme manifestations of Technē's Paradox. After Malthus, weapons rhetors could use similar, familiar rhetorical strategies to invent arguments for the preservative and destructive capacities of weapons. As a pre-text, Malthus thus provided the historical, conceptual, rhetorical, and embodied understanding of military technologies that is necessary for understanding the latter-day weapons advocacy that I analyze in subsequent chapters.

The *Principle of Population* became influential, winning many adherents among rival political economists and politicians with its vivid elucidation of how the unequal powers of rapid population growth and plodding advances in agricultural technologies affected poverty, emigration, war, and commerce. He was not the first writer to explore the vexed relationship between human populations and their means of subsistence, but he brought the debate out from obscure texts about political economy and into common conversation, the popular press, and the British Parliament.[3] As early as 1800, the book circulated among government officials, and as debate about whether to abolish or reform the Poor Laws dragged on, Malthus became something of a pundit.[4] He gained the ear of Samuel Whitbread, M.P., in 1807 and influenced the reforms advocated by both the 1817 Select Committee on the Poor Laws and the 1826 Select Committee on emigration.[5] He testified about emigration to the Select Committee of the House of Commons in 1827, and their final report showed a clear Malthusian influence.[6] Loath to admit the moral implications of eliminating large portions of the masses who clogged London's streets through forced parish resettlement, forced emigration, incarceration in work houses, and general abandonment, and when provoked to address the "principle of population" Members of Parliament considered solutions that drew from the darker implications of Malthus's arguments rather than from his exhortations for people to show ethical and procreative restraint.[7] In 1834, just after his death, the British government validated the "principle of population" and "Malthusianism" by passing the Poor Law Amendment Act, which presented paupers with an unappealing dilemma.[8] The law required them to choose between either banishing themselves to decrepit workhouses or refusing all future welfare.[9]

Notwithstanding his many protestations that he would never advocate unethical treatment of the poor, whether through violence or policy, the meaning of his work had spun out of Malthus's direct control, and Malthus the man transformed into Malthusianism the ideology.[10] For paupers and their advocates, it was Malthus's inflammatory writings that were "pernicious" and "very bad things," and not those of Malthus's critics. James Bonar, an early Malthus scholar and biographer, sarcastically proclaimed of his subject, "He was the 'best-abused man of the age.' Bonaparte himself was not a greater enemy of the species. Here was a man who defended small-pox, slavery, and child-murder; who denounced soup-kitchens, early marriage, and parish allowances."[11] But, as Kenneth Smith noted, if his critics "attack views which Malthus did not enunciate, but which were deduced with or without his consent from his work, this is quite a legitimate undertaking. . . . What men thought Malthus said was as important in the eyes of his critics as the very things he did say."[12] His infamy increased as his *Essay* circulated, and Malthusianism came to refer to harsh governmental treatment of the poor, ignoring the poor, punishing the poor, blaming the poor for their poverty, preventing the poor from procreating, and eliminating the poor.[13]

In addition to his influence on Poor Law reform and the formation of Malthusianism, Malthus gained fame as a political economist and protodemographer. In 1805, the East India College at Haileybury appointed him Professor of History and Political Economy to train aspiring colonial bureaucrats to work abroad for the East India Company. From his post as the first professor of political economy in Britain, he published pamphlets about prices, wages, rent, money, commodities, grain importation, the pedagogical importance of the East India College, and the treatises *Principles of Political Economy* (1820) and *Definitions in Political Economy* (1827). Among economists, the term "Malthusian" came to refer to theories of diminished economic returns derived from population growth, and basing wage theory on the "principle of population." From his economic research, Malthus recognized that history tended to focus on the wealthy, so his quest to prove his population theory by compiling as much statistical information about understudied poor people made him a founding figure of demography.[14] As Malthus published revised editions of the *Essay* in 1803, 1806, 1807, 1817, and 1826, it found readership, praise, and dissension in a wide range of political, economic, academic, journalistic, scientific, and social circles.

Given the influence of his treatise, one way to understand the *Essay* as a pre-text of weapons rhetoric is to consider how Malthus's population theory and conception of technology function as paradoxical social knowledge.[15] "Social knowledge," according to David Zarefsky, is the "general understandings and beliefs that characterize a political culture."[16] Such knowledge is not a static entity, but one "characterized by a state of 'potential' or incipience" and an "imperative for choice and action," in the words of Thomas Farrell.[17] And as Marouf Hasian noted, Malthus's population theory became a commonplace, "a part of the taken-for-granted scientific, moral, and legal wisdom of early nineteenth century England."[18] As social knowledge, Malthusianism remains so, thereby compelling choices and actions about how to handle the global effects of technology on large populations and on the understanding that the survival of the entire human species is at stake during times of war. Thus, the suggestion that Malthus's *Principle of Population* functions as a pre-text for Technē's Paradox is meant not only to articulate a particular person's thoughts about population, globalization, war, and political economy at the end of the eighteenth century in London, but also to articulate an enduring pattern of deliberation about weapons' effects on humanity's fate. After Malthus, to use Ian Hacking's assessment of the idea of the economy, "what was once visibly contingent feels like it has become part of the human mind."[19] Humanity is "the population bomb" waiting to go off.[20] And after Malthus, the Paradox that technologies, including weapons, might cause humanity's total destruction and total preservation was social knowledge that humanity had to confront.[21]

To establish the *Principle of Population* as a pre-text for Technē's Paradox, the next section takes a middle-way approach in order to examine Malthus's biopolitical ideas about population checks with respect to being in the presence of large standing armies and the violent masses' embodied rhetoric. Then, I analyze Malthus's writings about population in order to demonstrate how he used sublime statistical ratios, universal generalizations, and machine *antistasis* to bring humanity into confrontation with its two paradoxical, global, and technological fates. I conclude by reflecting on the suitability of Malthus's *Essay* as a pre-text for understanding subsequent weapons rhetoric.

1. Malthus's Principle and the Weight of So Many Soldiers

In order to establish a material perspective for understanding Malthus's rhetoric in the vein of the middle way between words about weapons and weapons themselves, this section examines the persuasive force of militarized human bodies—an instantiation of the population bomb. I focus on how the collective bodies marshaled for continental warfare in Europe during the Napoleonic Wars demonstrated, as Malthus would say, the activity of warfare as a potential "check" to population growth. War between the masses—both internal and external to states—appeared to corroborate nineteenth-century concepts of war that rationalized the annihilation of large populations in the name of preservation. Bodies motivate us to certain political beliefs and actions when amassed both with us and against us.[22] This line of thinking is known as "body rhetoric."[23] In one sense of the term, body rhetoric can refer to how "rhetoric is articulated through and by bodies."[24] In another sense of the term, bodies "make meaning."[25] Bodies also persuade in their display of injuries. And bodies injured by particular weapons lead us to make conclusions about the causes of the wounds, when injuring people "provides a record of its own activity."[26] But it is more fruitful for understanding Malthus as a pre-text for Technē's Paradox to consider body rhetoric as the ways that bodies motivate belief and action. Embodied rhetoric is especially observable when connected to other objects, discourses, lived bodies, and embodied practices in what T. Kenny Fountain called "object-body-environment intertwining."[27] I suggest that when human bodies are connected to other bodies in increasing levels of magnitude, the sheer number of bodies present in a given demographic group, in a given state, and in the entire world provides increasing levels of rhetorical weight. In this section, I first provide a more detailed account of Malthus's population theory. Then I examine what being in the presence of the collective and accumulating evidential weight of the militarized masses entailed, arguing that the rhetorical weight of "*an array of bodies*"[28]—massed soldiers, militarized civilians, and corpses—influenced the development of an embodied and biopolitical governmental consciousness that provided the material pre-text that is crucial for understanding the development of paradoxical weapons rhetoric.

At first glance, Malthus's "principle of population" might seem more straightforward than paradoxical. According to Malthus's basic argument, "The power of population is indefinitely greater than the power in the earth to produce subsistence for man. Population, when unchecked, increases in a geometrical ratio [1, 2, 4, 8, 16, 32, etc.]. Subsistence increases only in an arithmetical ratio [1, 2, 3, 4, 5, 6, etc.]. A slight acquaintance with numbers will shew the immensity of the first power in comparison to the second. By that law of our nature which makes food necessary to the life of man, the effects of these two unequal powers must be kept equal."[29] Few could assail the basic premise that humans procreate more easily than human ingenuity can increase the means of subsistence. As publisher Hutches Trower wrote to Malthus's friend and rival political economist David Ricardo in 1821, "Whether population will double itself in 25 or in 50 years is of no moment as far as the *principle* is concerned."[30] Trower's point was that whenever population increases, feeding all of the additional people is onerous. And extrapolating the effects of Malthus's ratios showed that preserving one population entailed threatening other populations. In the controversy surrounding what to do about poor people, industrialization, and popular ethics, policy debates often took paradoxical turns as people weighed the advantages and disadvantages of population management.

Management of the population bomb underpins the Malthusian link to Michel Foucault's concept of biopolitics. Biopolitics, or "the calculated management of life" with "population controls," involves an array of "mechanisms" that tend to regulate large populations according to Malthusian principles, such as modifying and lowering the mortality rate, increasing life expectancy, stimulating the birth rate, using "regulatory mechanisms" to ensure a homeostatic population, and ensuring compliance by way of "security mechanisms."[31] Biopolitics is a useful frame with which to understand the body rhetoric of Malthus's time for several reasons. First, Foucault dated the "liberalism" that underpinned the emergence of biopolitical thought to Malthus's scene—England at "the end of the eighteenth and in the first half of the nineteenth century."[32] Second, Malthus's central concern—population—was the "central core" of Foucault's mid-1970s lectures about biopolitics.[33] Third, Foucault proposed that biopolitics emerged out of Malthusian-style demography, or "the evaluation of the relationship between resources and inhabitants," and the capitalistic regulatory "adjustment of the phenomena of population to economic processes" for the "global mass."[34] In these ways, "Malthus described what Foucault might call a biopolitical History."[35] Keeping the "two unequal powers" of procreation and agriculture equal requires both massive technological projects and governmental oversight in order to abrogate the universal and perpetual natural laws that Malthus proclaimed in his *Essay*. By arguing that the benefits of managing sustenance and procreation would entail making choices that would harm a large segment of England's population—the poor—in order to preserve the remaining population, Malthus announced the prototypical importance of Technē's Paradox for deliberations about how to use technologies. And just as Malthusian population management involves the maintenance of life and death for large populations, so, too, does biopolitics attend to living populations and, with the counterpart principle of "thanatopolitics," their demise.[36] Thanatopolitical policies (letting die and allowing to be killed) are the counterpart to the biopolitical endeavor to preserve and multiply human lives.[37]

"Positive checks"—a fundamental aspect of Malthus's population theory that incorporates both the positive check of war and the collective weight of enormous standing armies—corroborated the Malthusian lesson that violence threatened certain populations with total annihilation while preserving others. According to Malthus, "positive checks" bring population and sustenance into balance with each other by killing the excess population. Positive checks do not work alone to regulate population levels; they work in tandem with "preventative checks," which are the prudence and "moral restraint" that people use when deciding whether to marry and procreate.[38] Preventative checks are an essential component of Malthus's population theory, but they are nontechnological in character. In contrast, positive checks are rooted in whether a population has the technological capacity to assert control over procreation and death. The number of living and dead bodies demonstrates the level of population control a state achieves.[39] Malthus listed numerous positive checks: "unwholesome occupations, severe labour and exposure to the seasons, extreme poverty, bad nursing of children, great towns, excesses of all kinds, the whole train of common diseases and epidemics, wars, pestilence, plague, and famine."[40] That Malthus lumped technological phenomena—factories, cities, and the means to wage war—together with more organic phenomena (e.g., breast feeding, diseases, and harsh weather) as external agents of population decline implied that population management required technical control over all biological and behavioral aspects of human life in a quest to master nature.[41] The more that humanity can use technologies to control these external agents of population decline, the more it can preserve life. But, as Malthus always reminded his readers, the preservation of life comes at the severe cost of disaster once a population reaches the "Malthusian limit."[42] If not left to run their "natural" course, then oversight of "positive checks" is required to avoid both massive die-offs and catastrophic population explosions.

While Malthus was revising the *Essay*, the states fighting in the Napoleonic Wars were amassing and killing unprecedented numbers of soldiers. And if "*presence* acts directly on our sensibility," as Perelman and Olbrechts-Tyteca argued, then the soldiers' collective body mass gave rhetorical presence to the possibility that one large population could exterminate another for the sake of self-preservation, even at the cost of one's own population.[43] The presence of so many corpses and soldier-citizens provided the embodied rhetorical counterpart to biopolitics. According to Foucault, in the early nineteenth century, "Wars are no longer waged in the name of a sovereign who must be defended; they are waged on behalf of the existence of everyone; entire populations are mobilized for the purpose of wholesale slaughter in the name of life necessity: massacres have become vital."[44] Bonaparte projected this logic and this massed-body rhetoric in his words and on his battlefields. "A man like me does not give a shit about the lives of a million men," he said of the death and destruction he orchestrated.[45] The amassed corpses and soldiers of the Napoleonic Wars may not have impressed Bonaparte's ethical sensibility, but population growth empowered ruling bodies to wage war and "made wars more possible" simply because there were more bodies available to fight and die.[46] Enlarging the scope of warfare to incorporate the masses meant greater interdependence among industrial production, armaments, and the standing armies of England, Austria, Germany, and France.[47]

With weaponry and standing armies increasing at a pace closer to Malthus's geometrical rate than his arithmetical rate, the context of the French Revolution and Napoleonic Wars lent another level of embodied credence to Malthus's generalization that terrible violence would erupt from global population growth. France's *levée en masse* of 1793 aimed for the total mobilization of France's citizenry, and military strategy adapted to use the increasing populations for war. Before the *levée*, for the most part, typical citizens were not targets of lawful warfare.[48] But following Edmund Dubois-Crancé's 1789 dictum that "in France every citizen must be a soldier, and every soldier must be a citizen, or we will never have a constitution," the *levée* legislated universal militarization. It decreed that "all Frenchmen," including women, children, and the elderly, "are permanently requisitioned for service into the armies," either to fight or provide logistical support.[49] Thus, in its attempt to mobilize the entire

French population of thirty million citizens, the *levée* marked a signature moment.

The *levée* meant that not only would all French citizens become citizen-soldiers, but also that, by embodied rhetorical compulsion, so, too, would all foreign citizens and any other population deemed an enemy of France.[50] And as "the exterminating angel of liberty" caused increasing destruction, a nascent concept of "total war" that called for the complete annihilation of enemies began forming.[51] In total-war logic, which developed from German military strategist Carl von Clausewitz's idea of "absolute war," defeating an enemy state requires destroying its military capability, its economy, its citizens, and any other contributing factor to state power. Rather than providing a limiting principle as prior war strategies had endeavored to do, "all limits disappeared in the vigor and enthusiasm shown by governments and their subjects."[52] Napoleon's army, for one, seemed unlimited in its size and its capacity for violence. Armed by a gunpowder factory that produced thirty thousand pounds a day, and by a rifle industry that produced 750 a day, the French Army—at an unprecedented strength of 750,000 citizen-soldiers—could swarm anywhere in Europe like no prior army could.[53] In the Vendée region, for instance, "the rhetoric of total war was fully translated into blood-streaked, exterminatory fact."[54] French troops proceeded to massacre enemies in Cairo, Jaffa, Haiti, and Spain, and nineteenth-century colonial expansion exported the threat of large standing armies around the world.[55]

The militarized masses were a population bomb not just in terms of their procreative power, but also in terms of their power as weaponized bodies. The body rhetoric of massed soldiers materialized in Malthus's writings as he recognized the usefulness of hungry, jobless people for military conscription. Malthus depicted how the powerless statistical fringes of overpopulated societies empowered states to see them as the means to wage war. He wrote, "The ambition of princes would want instruments of destruction, if the distresses of the lower classes did not drive them under their standards. A recruiting serjeant always prays for a bad harvest and a want of employment, or, in other words, a redundant population."[56] New weapons were unnecessary when the redundant population made such good cannon fodder and when "Europe was involved in a most extensive scene of warfare requiring all its population."[57] The possibility of an internal total war of revolt was more frightening to Malthus, though, than a French invasion was. Some workers, such as the Luddites, rebelled by destroying the machines that replaced them in factories, and others such as the journalists of the "pauper press"—the radical, working-class journalism that flourished in the 1820s and 1830s—sought to compel the revolutionary massacre imagined by Malthus.[58] In a letter to Ricardo in 1819, Malthus asserted that a British revolution would be much more violent than France's, and the "massacre would in my opinion go on till it was stopt by a military despotism."[59] However, the British government had responded to the Luddites with the Frame Breaking Act of 1812 that made machine breaking punishable by death, and Malthus credited the British standing army with holding revolution in check.[60] The presence of amassed military bodies and regulatory thanatopolitics thus stifled British revolutionary ambitions.

The embodied rhetorical presence of so many soldiers—alive and dead—made the willful, destructive use of weapons a crucial variable for thinking about the problems of population management. This rhetorical embodiment, when felt alongside the reading of Malthus's *Essay*, lent the gravity of his population theory a sense of urgency. Enlarging standing armies and arming them with mass-produced weapons promoted like behavior in rival states, while Malthusian logic seemed to justify the violence of governments that promoted state self-preservation.[61] At "society's 'threshold of modernity'"—at Malthus's time—"the life of the species is wagered on its own political strategies," wrote Foucault.[62] The wager meant that being in the presence of massed and armed rival populations entailed a response. "To preserve life, it became necessary for governments to "expose one's own citizens to war, and let them be killed by the million," Foucault claimed.[63] Giorgio Agamben noted that the study of demography, in tandem with biopolitics, granted governments the capacity to allow their own populations to become the victims of state-sanctioned holocausts.[64] Weapons, as a component of the biopolitical regimentation that was needed to avoid internal conflict that was modeled on the French Revolution, were, in the words of Foucault, components of "a technology which aims to establish a sort of homeostasis, not by training individuals, but by achieving an overall equilibrium that protects the security of the whole from internal dangers."[65] For the longitudinal history of weapons rhetoric, Malthusian biopolitical and thanatopolitical powers—enhanced by the presence of weapons in so many hands and the threat of annihilation that they posed for so many people—bolstered racist thinking and Nazism, and then culminated in Cold War–era nuclear deterrence when "the biological existence of a population" was "at stake," which is a key focus of chapter 4.[66]

In a biopolitical twist of body rhetoric, the standing armies that arose from the Napoleonic Wars ended up not checking population growth but spurring it. Malthus registered surprise that, at the end of the Napoleonic Wars, Great Britain's wealth and population had increased at "a more rapid rate than was ever experienced before."[67] Similar data showed that France's population increased during the Revolution so much that Malthus conceded "considerable surprise" that the conflict resulted in an "undiminished state of the [French] population in spite of the losses sustained during so long and destructive a contest."[68] With continental war as the context, more people were needed to fight, defend, and die. The population bomb grew more powerful.

Thus, the body rhetoric of massed soldiers, their weapons, and their corpses worked in tandem with Malthus's words to make his population theory appear plausible. But it was not only war technologies that preserved and destroyed. Apparent technological necessities for human survival—such as improvements in agriculture, sanitation, medicine, and transportation—could appear bad because they kept too many people alive. For instance, the presence of more than enough food signaled that a crisis was coming, and such was Malthus's estimation in 1809 of the Irish poor's reliance on potatoes as an agricultural staple, a reliance Malthus predicted would lead to famine.[69] This Malthusian extension of the "principle of population" shows how fundamental Technē's Paradox is to technology in general: controlling disease, epidemics, pestilence, and plague with improvements in medicine and sanitation helped to sustain poor people, but such inventions kept them alive to face death in other gruesome ways. Therefore, beneficial technologies needed to be further controlled, mastered, and manipulated in order to control, master, and manipulate ever-larger populations. In terms of Technē's Paradox, Malthusianism functioned as a guiding "Spirit, or Consciousness" that placed technologies at the center of human affairs, and as the lived material relations that we experience when an idea "becomes a universal self-developing organism."[70] However, the relevance of Malthusian checks to later weapons rhetoric also derives from the rhetorical tactics he used to expand his argument that some populations will be driven to the point of extinction in order to preserve the biological survival of other populations.

2. The Mathematical Sublime, Universal Generalizations, and Machine *Antistasis*

As Malthus's fame and infamy increased, his writings goaded people to think about the relationship between technology and humanity's fate, and his words helped to establish the conceptual and material relationship between Technē's Paradox and weapons. When Malthus speculated about global human extinction, three rhetorical tactics helped formulate Malthusianism as social knowledge that extended the theoretical effects of weapons to make the prospects of total human destruction and preservation seem possible. First, he amplified the possible effects of his natural law with sublime statistics—the geometrical and arithmetical ratios—that depicted the future with absurd, improbable environmental scenarios. Second, he amplified local technological effects with generalizations that made Technē's Paradox appear universal. And third, he amplified the concrete aspects of technology with a type of *antistasis* in which the word "machine" revealed how single devices can become imbued with so much simultaneously preservative and destructive power. As these rhetorical tactics toggle readers' minds back and forth between thinking about the utopian preservation of all people and their dystopian total destruction, Malthus's *Principle of Population* provides the pre-textual significance needed for understanding paradoxical weapons rhetoric.

Malthus proffered the geometrical (1, 2, 4, 8, 16, 32, etc.) and arithmetical (2, 3, 4, 5, 6, etc.) ratios as statistical proofs that humanity's capacity to increase in population far outpaces its capacity to increase agriculture and subsistence.[71] Should the ratios hold true, then human populations would soon reach unsustainable levels, and the numerical disparity between the ratios indicates how Technē's Paradox became a pervasive conundrum with respect to population control. The *Essay* used the ratios to describe the implausibility of American and British abilities to increase agricultural outputs to meet a speculative spike in population, and the resulting calculations amplified the probable disparity to the point of sublimity.

In America, Malthus found corroboration that population doubles every twenty-five years when abundance permits. So, according to the ratios, population would outpace food production in three centuries by a ratio of 4,096 to 13, "and in two thousand years the difference would be almost incalculable, though the produce in that time would have increased to an immense extent."[72] The idea that the population would surpass the means of subsistence by the ratio of 4,096 to 13 in three centuries might sound "absurd," but such ratios offered a clear illumination of the potentially devastating effects produced at the Malthusian limit.[73] Catastrophe would ensue long before the ratios could increase to 4,096 to 13. "In these ages want would be indeed triumphant, and rapine and murder must reign at large," he wrote.[74] Malthus conceded that this extreme disparity of 4,096 to 13 "could never have existed," but the stunning character of the ratios left the fate of the American population open to gloomy speculation.[75]

Along with the impossible statistical disparities that Malthus calculated from his ratios, he used a type of "mathematical sublime" to prove the "principle of population."[76] Kenneth Smith called the ratios "bewildering," indicating their connection to Technē's Paradox. Smith wrote that "what Malthus does . . . is to make men's minds reel in an attempt to reach infinity by counting," and that as the numbers become more and more implausible, they "serve to bewilder the reader; they serve to impress him with the overwhelming superiority of the power of population."[77] Beyond bewilderment, Malthus's ratios provided readers of the *Essay* with a stunning, overwhelming, and fearful sign of the Malthusian limit by amplifying the problem of overpopulation to the point of sublimity. In this way, the rhetorical function of the ratios derived from Malthus's use of what Kant called in 1790 "the mathematical sublime." According to Kant, when the magnitude of numbers grows more and more enormous, "our imagination strives to progress toward infinity, while our reason demands absolute totality as a real idea, and so [the imagination], our power of estimating the magnitude of things in the world of sense, is inadequate to that idea."[78] As the disparity between procreation and agriculture grows in magnitude and culminates in a ratio like 4,096 to 13, it becomes more difficult to grasp the total severity of the problem. The numbers are so large that they both draw attention and defy simple assessment. Judging Malthusian ratios, in the words of Kant, thereby "strains the imagination (of expansion) to its limit."[79] The mathematical sublime thus inhibits understanding the enormity of the total claim. Malthus's amplification of the ratios to the point of sublimity perhaps forced his audience to consider the overwhelming direness of the population problem, while a lesser disparity might have indicated that technology could be implemented to maintain a balance between the two competing powers of productivity before disaster strikes. Yet even in its purported impossibility, the ratio demanded attention by amplifying why populations must control their numbers with technologies.

Malthus's ambivalence regarding whether the ratios should be considered exact calculations further indicates that they helped to prove his population theory more by producing a sublime impression than by producing a rigorous statistical proof. Malthus emphasized the ratios in the early editions of the *Essay*, but he qualified his position on the value of "facts and calculations" in a later edition, writing that "should any of them nevertheless turn out to be false, the reader will see that they will not materially affect the general tenour of the reasoning."[80] Malthus used the ratios as static, ideal representations that rigidified the chaos of life into a natural law. They represented true premises, but the exact numbers were false in most cases. Numbers and facts could vary without impugning the universality of the "principle of population," because these inconsistencies did not disprove the "principle" per se.

The example of Britain provided Malthus with another opportunity to amplify the effects of the population bomb to the point of sublimity, this time by using the geometrical ratio, instead of the arithmetical ratio, to demonstrate agricultural limits. By imagining agricultural innovation as proceeding at a geometrical clip, Malthus, for a moment, adopted a position contrary to his own ratio.[81] Malthus conceded that Britain might geometrically double its agricultural output in the upcoming quarter century as the maximum possible increase.[82] But Malthus averred that any rapid initial geometrical increase in agricultural productivity would be brief and unrepeatable. He argued that a further geometrical doubling of the land's agriculture in the next quarter century (a quadrupling of the original productivity) would be "impossible."[83] Each time Malthus multiplied agricultural increases by the geometrical rate, the prospective ingenuity required to reach such feats approached the sublime. He therefore moved closer and closer back to his original position that agriculture increases in an arithmetical rate at each calculation. Britain did not possess the technological ingenuity needed to quadruple agriculture in fifty years, much less to pull off the necessary eight-fold increase in seventy-five years, and so forth. In a time just after a string of famines, geometrical doublings of agricultural productivity sustained over many decades would not have seemed feasible in its almost unimaginable depiction of abundance. Malthus indicated this sublimity with sarcasm. "In a few centuries it would make every acre of land in the Island like a garden," he mused.[84] He directed this cutting remark, and his British example as a whole, at political philosopher William Godwin's conception of anarchy as a leisure-filled utopia. But at the same time, Malthus's sarcasm amplified the limits of agriculture's preservative power. The conversion of Britain into an uninterrupted lush garden was an unattainable endeavor that would have required a mathematical increase in subsistence as sublime as his statistical predictions about America, so Malthus intimated that practical attention to Britain's overpopulation problem should not rely on agriculture alone. Rather, Malthus argued, the solution would derive from moral education and elimination of the

Poor Laws, while Malthusians argued that the solution would derive from the accumulation of better technologies for population control.

Once Malthus had portrayed the conversion of Britain's entire surface area into a massive garden as preposterous, he again used his ratios to depict a more "melancholy picture" of Britain.[85] If agriculture would not preserve everyone, then it would destroy many. He forecasted that in a few cycles of statistical progression, "we shall see twenty-eight millions of human beings without the means of support; and before the conclusion of the first century, the population would be one hundred and twelve millions, and the food only sufficient for thirty-five millions, leaving seventy-seven millions unprovided for."[86] The Malthusian limit marked a boundary between those welcome at "nature's mighty feast" and those shunned—the seventy-seven million citizens who would be struggling for basic survival.[87] This boundary, enumerated by the ratios, cast out a certain percentage of the population incapable of providing its own sustenance by portraying them as unnecessary at best and a threat at worst.[88] Malthus's prediction of so many deaths once again approached the mathematical sublime. The doomed population, seventy-seven million strong, thus defined those members of a population who most needed technological preservation in the face of probable annihilation.

When the statistical disparity between the powers of procreation and agriculture were so extreme, the ratios legitimated the redundant population's marginalization. Techniques for controlling the redundant population did not always focus on preservation, but often focused instead on abandonment or "checking" their growth with the biopolitical regulatory power of "letting die."[89] Most cases of overpopulation threatened the already poorer and powerless classes with starvation, conscription, and "unwholesome" jobs, so the British government, industries, and military could continue or increase their ill treatment of them with impunity. Neither agriculture nor manufacturing could provide the redundant population with subsistence, so they were destined not to lives in an abundant garden, but to lives, in the words of Thomas Hobbes, "poore, nasty, brutish, and short."[90] Malthusians, if not Malthus, thus regarded the poor's superfluity and societal burden as a justification for their harsh and brutal treatment. Their sublime speculative numbers warranted checking the population before disaster might strike. The technology of the workhouse was not meant to preserve Britain's statistical remainder but to eliminate them through profitable attrition and procreative deterrence.

In later editions of the *Essay*, Malthus revised some of his statistics in order to reflect new evidence without changing any of his claims, and he finally dropped the ratios from his *Encyclopedia Britannica* essay on population and *A Summary View of Population*.[91] But Malthus did not drop the competing productive premises that he intended the ratios to prove. Even without amplifying his ratios to the point of sublimity in the final edition of the *Essay*, his population principle still drew attention to Technē's Paradox. The statistics made the population problem appear sublime, but after their removal, the principles behind the ratios still implicated technology as the solution and the cause of humanity's precarious relationship to its resources. In the decades after Malthus's death, agricultural innovations did keep pace with population growth much better than Malthus anticipated they would. Notwithstanding the Irish potato famine, the population bomb exploded with many fewer shortages and famines than predicted. Malthus's population theory, however, showed that large populations should not find solace in agricultural abundance, and it warned that the larger the global population grows, the greater the eventual catastrophe will be.

Part of the mathematical sublimity inherent in imagining the outcome of Malthus's ratios involves discerning the speculative terrible effects that massive disparities between population and subsistence will cause. According to Malthus, any large disparity between population and subsistence would cause catastrophic warfare long before it reached 4,096 to 13 or left "seventy-seven millions" to starve. Violence would become necessary in order to assure basic survival, and the greater the disparity, the greater the magnitude of biblically apocalyptic violence.[92] In turn, Malthus tended to generalize the violent effects of his ratios to the entirety of the earth in order to prove the universality of his "principle of population." His ratios worked in concert with his generalizations: once the ratios established war as the probable result of the "principle of population," then his generalizations amplified the scope and severity of war to a global scale. Therefore, when Malthus incorporated war into his universal principle, war became a universal threat. Malthus thus amplified his generalization to the point of universal abstraction, far beyond what he could prove from localized examples. If every society must abide by the "principle of population," then Malthus could promote any one local effect of his population theory as the speculative fate of the whole earth as long as it fit the general tenor of his "principle." By moving from localized examples to generalized abstraction, though, Malthus demonstrated increasing levels of violence. Malthus, after all, lived in an era when such doomsday generalizations gained traction.[93] Within this zeitgeist, specific technologies appeared to be poised to annihilate all of humanity in the same ways that they annihilated local populations in the past, and Malthus chose to amplify the probability of horrific warfare: agriculture will feed humanity to the brink of catastrophe, and weaponry will facilitate the ensuing collapse. He amplified his generalizations about war in several steps that each increased the level of magnitude: he implied global annihilation by using hyperbolic language to depict a singular, ancient conflict; he made a specific inductive argument that moved from an American example to a universal conclusion; and he amplified his generalization about global bloodshed to the point of abstraction.

When Malthus moved from analyzing specific examples of his ratios, such as events in America and Britain, to speculating about the fate of all people, his population theory underwent globalized generalization.[94] After examining some effects of British emigration on population change, he made such a generalization. He wrote, "To make the argument more general and less interrupted by the partial views of emigration, let us take the whole earth, instead of one spot, and suppose that the restraints to population were universally removed."[95] In addition to Britain, Malthus proved the basic validity of the premise that population levels are tied to sustenance by spanning the globe to find examples. Especially in the later editions of the *Essay*, Malthus justified his generalizations with the sheer evidential weight of many geographic examples. As his demographic studies traversed the globe in time and space—from ancient times to modern, from England to Europe, and from Europe to America, Africa, East Asia, Southeast Asia, Siberia, the South Seas Islands, and beyond—from region to region, he ended up transcending it. The "whole earth" proved his premises correct and established his population theory as "a universal law, valid for all populations at all times."[96] Societies attempt to use different techniques to manage population and agriculture with varying degrees of success and failure, and no society can escape the "incontrovertible truths" that abundance empowers procreation, famines check procreation, and unchanging productivity keeps population growth somewhat stable.[97] Convincing his audience of the universality of this balance was perhaps Malthus's greatest rhetorical achievement.[98]

Malthus also applied his generalizations to warfare. In the first *Essay*, before making an explicit "whole earth" generalization about the effects of war, Malthus amplified the level of violence in a small conflict in order to imply a more general global catastrophe. Looking backward to an ancient conflict, the "barbarian" conquest of Scythian shepherds, he wrote, "Gathering fresh darkness and terror as they rolled on, the congregated bodies at length obscured the sun of Italy and sunk the whole world in universal night. These tremendous effects, so long and so deeply felt throughout the fairest portions of the earth, may be traced to the simple cause of superior power of population to the means of subsistence."[99] Malthus then reflected that "in these savage contests many tribes must have been utterly exterminated."[100] This "utter extermination," whether hyperbolic or not, provided a concrete example of one population that destroyed another population in a war caused by the dictates of biological survival. It was an example that Malthus could use to imagine a larger catastrophe. Knowing that the Scythians were exterminated to make room for ancient colonists, the fate of an overpopulated planet appeared even bleaker. It stood to reason that modern indigenous populations would meet the same fate while European colonies proliferated, according to Malthus's universal generalizations. And by implication, overpopulation threatened the whole world. But the magnitude of the "tremendous effects" suffered long ago by the Scythians would be much greater at the turn of the eighteenth century, because the mass of "congregated bodies" would be much larger, the potential "darkness and terror" greater, and more of the "whole earth" than just Italy affected. Thus, he amplified the ancient Scythian crisis to portray a modern problem of universal importance.

Malthus made a more specific inductive argument that generalized from specific examples of conflicts in America to amplify a universal conclusion that the global population could suffer a brutal war of extermination. He prophesied that "if America continue increasing, which she certainly will do, though not with the same rapidity as formerly, the Indians will be driven further and further back into the country, till the whole race is ultimately exterminated."[101] Agricultural improvements did not provide liberty to the entire American population and failed to provide moral protection. Domestic expansion of crops and manufactured goods came at the price of First Nations people lives. When European colonists proliferated, they wiped out entire indigenous populations, counterbalancing sustenance and population with "positive checks."[102] Malthus took his empirical observations of violence against First Nations people and the universality of his population theory as an impetus to generalize extermination and amplify a possible global crisis. Malthus wrote, "These observations [of indigenous wars] are, in a degree, applicable to all the parts of the earth, where the soil is imperfectly cultivated."[103] But according to Malthus's skeptical view of agriculture's capacity to keep pace with population growth,[104] he was asking his readers to imagine a scenario in which every population in every country across the globe would die in an unmerciful, horrific spate of killing. He was asking his readers to imagine a slaughter in which women and children are wiped out by invaders who are driven to commit atrocities by the simple dictate to survive. The result of such a war of extermination in "all parts of the earth" would jeopardize the survival of large populations on every continent, if it did not result in the near-extinction of humanity. The universality of the "principle of population" implied to readers that they, too, might end up committing atrocities in a time of crisis or fall victim to someone else's hatchet or bayonet.

Having generalized "utter extermination" from the local level to the global level, Malthus further amplified his generalization of universal warfare to the point of abstraction. He described how overpopulation and war combined to push the level of violence upward in an increasing scale of annihilation: "Premature death must in some shape or other visit the human race. The vices of mankind are active and able ministers of depopulation. They are the precursors in the great army of destruction; and often finish the dreadful work themselves. But should they fail in this war of extermination, sickly seasons, epidemics, pestilence, and plague, advance in terrific array, and sweep off their thousands and ten thousands. Should success be still incomplete, gigantic inevitable famine stalks in the rear, and with one mighty blow levels the population with the food of the world."[105] Malthus imagined the result of such a population explosion as a depraved bloodbath and food as the final weapon. Unleashed within the framework of early capitalist empire building, "evil" reigns victorious over humanity's ability to manage population levels.[106] Malthus implied that, at the climax of this progression from "depopulation" to "war of extermination" and then to "gigantic inevitable famine," the population bomb threatens humanity with extinction.[107] The number of dead would accumulate as vices, war, weather, and various types of diseases each kill tens of thousands. Malthus portrayed the most destructive positive check, famine, as capable of "completing" any mass death by providing the maximum number of bodies needed to recalibrate procreation with agriculture, a number that might be well beyond tens of thousands.[108] He extrapolated catastrophic population crises not with one primary cause of death but with all of them operating in concert at a magnitude theretofore unknown. This catastrophic climax was an abstraction, in part, as he did note that population would "level" after all of the death and destruction, but this image of war made the catastrophe sound much direr than normal, more specific population readjustments. Furthermore, when Malthus established the universality of the catastrophe's cause, he implied that this accumulated disaster was both repeatable and plausible across the planet and capable of unleashing unheard-of and improvable levels of violence: a global overpopulation crisis could level the population at zero. He thus abstracted his "principle" by speculating about global destruction on a scale he could not exemplify. Basing his generalizations on empirical fact would have meant that the "end" had already come.

Malthus repeated his warning "that the general principles on these subjects ought not to be pushed too far," and that concerned citizens should not harm the poor for the sake of "remote consequence."[109] But Malthus always pointed out the remote consequence, which tended to sound much more terrible than any misery arising from short-term population management. Malthusians therefore extrapolated a justification for brutal treatment of the poor from arguments that Malthus had made explicit in his treatises. Generalized violence and war, as primary negative consequences of his "universal law," appeared to be inevitable everywhere.

Thus, the importance of Malthus's amplified generalizations about the population bomb to weapons rhetoric derives from the way Malthus depicted the entire human race as technologically and militarily imperiled. When Malthus amplified local conflicts into a speculative global conflict, he implied that, as a correlate to war, weapons came to represent an amplified and universal threat as well. Paradoxically, by providing the means to check enemy populations, weapons also came to represent a fortunate and beneficial way to recalibrate population with the means of survival. Evolutionary theories from the likes of Charles Darwin and Alfred Russell Wallace, as generalizations of Malthus's generalization, began to make sense in their mimicry of the natural scientific method of Isaac Newton.[110] The two biologists used Malthusian inductive generalizations to argue that the history of specific species' populations proved survival behaviors and tendencies for the entire biological world.[111] The later development of social Darwinism and its survival of the fittest rationale, in which humanity was the endangered species, indicates how Malthusian generalizations about the fate of humanity radiated outward to influence political rationalizations for innovating, adopting, and deploying more and more powerful weapons. Malthus thus demonstrated that the preservation of humanity would rely on the survival of merciless wars over resources, and this generalization became a rhetorical commonplace that weapons rhetors could draw upon to argue that specific weapons could cause total world destruction or preservation. This imperiled collective fate became an undergirding rationalization for advocating and resisting weapons of mass destruction. The survival of the fittest in times of total war demanded innovation of the most destructive weapon. In this way, Malthus's amplification of warfare to the point of universal abstraction presented the violent interplay of technologies and populations as an insurmountable "natural law."

The provocative ingenuity observable in Malthus's statistics and generalizations carried over to how his use of commonplace machine metaphors proliferated ways of understanding the relationship of technologies to society. Machine, in its multiple meanings, functioned in his population theory as a type of *antistasis*, or "repetition of a word in a different sense" for which each usage alludes to different connotations of the term.[112] Malthus used a literal meaning of machine when he described the negative effects of "unwholesome" factory work on the poor, and he used a metaphorical meaning of machine to refer to the capitalistic functioning of government. In this way, machine was polysemous. Although he disparaged "unwholesome manufactures" in his most specific engagement with technologies, his attitude toward industrial technologies remained ambivalent as his overall assessment of machines became more positive. Malthus's switching between literal and metaphorical machines revealed an undergirding logic of industrialization: despite the problems of dangerous factories, human redundancy, and weapons production, capitalism and its machines improved society, so they should be tweaked rather than overthrown. Machines, in the metaphorical sense of government and capitalism, made technology appear to be a preservative boon, but when machines materialized Malthus's abstract population theory in a destructive, nonmetaphorical sense for the masses, machines materialized Technē's Paradox.[113] The polysemy of machine *antistasis* vaunted the simultaneously destructive and preservative powers of capitalist technologies, prepared Malthus's audience to be capable of swapping a variety of technologies in and out of Malthus's argument, and empowered rhetors to insert his argument into almost any technological context, further proliferating Technē's Paradox.

Malthus used the machine metaphor for government in response to his rival, Godwin, so understanding Godwin's usage is essential for understanding Malthus.[114] By no means did Malthus or Godwin invent the mechanistic metaphor for government.[115] It was a banal commonplace, with perhaps the most famous prior example being Hobbes's "artificial man"—the Leviathan—which moved by engines, springs, and wheels that condensed humanity, technology, and governance into one complex metaphor that drove his vision of monarchy.[116] Machine metaphors can explain a state by treating it as a whole entity or by disassembling it into parts and thereby depicting the efficient inner workings of government and any other nonmechanical process meant to appear complex, automatic, and beneficial. In Godwin's case, he used machine *antistasis* to elevate the advantageous status of the machine in his philosophical paradigm. In a more literal sense, machines would support humanity's leisure by performing all of its toil and labor and, as a metaphor, machines connoted the smooth operation of a peaceful society based on universal liberation. Godwin rejected antitechnological attitudes as "uninformed and timid" as he described how to disassemble and reinvent the governmental mechanism.[117] He wrote, "The progress of science and intellectual cultivation, in some degree, resembles the taking to pieces a disordered machine, with a purpose, by reconstructing it, of enhancing its value."[118] Godwin's machine metaphor for government was less vexed by dichotomous meanings than Malthus's. The completed anarchist machine, fashioned from "the confused heap of pins and wheels that are laid aside at random" from the dismantled government, would run with precision like a machine and with machines.[119] Machines both worked in Godwin's utopia and symbolized how utopias would work as a society, and hence the metaphor complemented his politics.

Despite his political and technological disagreements with Godwin, Malthus used the machine metaphor for government without compunction.[120] To question benevolence as Godwin's "main-spring" of government, Malthus proposed disassembling the British government into parts in order to contemplate reforms that might strike a balance between population and sustenance. According to Malthus, "a society constituted according to the most beautiful form that imagination can conceive, with benevolence for its moving principle," will always revert back to an economic class system "with self-love the main-spring of the great machine."[121] The anarchist "main-spring" would malfunction; benevolence would break down within Godwin's political machine, wrecking it. Malthus therefore argued that rather than make order out of disorder, benevolence would make disorder out of order. A "main-spring of self love" might cause problems, but it would not break the entire government like, in Malthus's estimation, benevolence would.

If the design fails, then society must retool the mechanism. For Godwin's anarchism, this entailed a complete disassembly and reassembly, to the point of eliminating the state altogether and keeping only its "main-spring" in operation. For Malthus, the "main-spring," or "self-love," of the capitalistic "great machine" must be retained.[122] Malthus wrote, "The structure of society, in its great features, will probably always remain unchanged. We have every reason to believe that it will always consist of a class of proprietors and a class of labourers; but the condition of each, and the proportion which they bear to each other, may be so altered as greatly to improve the harmony and beauty of the whole."[123] In addition to demonstrating his political moderate-conservatism and antirevolutionary attitude, this passage provides an alternative prospect for poverty-stricken, redundant populations. With proper management on governmental and scientific levels and proper self-management on the moral level, redundant populations could be eliminated not through destruction, but by personal inhibition and preventative birth control.[124] This combination of techniques could save the masses from living in misery and thereby save the state as well. Individualism is thus inherent to maintaining the "great machine" of industrial capitalism.

Malthus's use of the machine metaphor vaunted capital as a preservative force, for both individuals and society. Industrial capitalism provided, according to Malthus, two important benefits to the entire British population—greater civil liberty for the individual and greater wealth and productivity for the state. Upon the introduction of new factory machines, "liberty came in their train."[125] However, liberty was power not for the poor, but for the state. Malthus made the economic benefits of machinery to the state most clear in *Principles of Political Economy*, in which he wrote, "Like the fertility of land, the production of good machinery confers a prodigious power of production."[126] Producing more liberty and wealth for the middle and upper classes meant the working classes improved only in the sense that they, more and more, worked in the dangerous occupations that garnered wealth for the owning class and stabilized the government by consolidating economic power. Liberty, for workers, entailed their capacity to exchange their labor for wages, rather than expecting entitlements, which may or may not come, from landed gentry and the government. As a devotee of Adam Smith, Malthus equated the freedom of the poor with the greater freedom of markets. And poor individuals were each free to turn the "main-spring" with their "self-love," but they remained powerless to do much else within the "great machine's" economy. The machine metaphor simplified the complex predicament of the poor in relationship to the state because it construed everybody, from the most destitute pauper to the wealthiest capitalist, as economically free individuals who could alter their situations by turning the self-love main-spring of capitalism. They could capitalistically preserve themselves from capitalistic destruction.

A few pages after his metaphorical use of "machine," Malthus used the term in its literal sense to refer to mechanical devices. Malthus focused his attention on manufacturing machines, but by associating the term "machine" with an array of technologies as he speculated about their beneficial and detrimental effects, his *antistasis* "invited"[127] readers to swap various machines in and out of his examination of factory machinery in order to understand and critique the ethics of capitalism better. Wondering what ancient Greeks would think of watches and telescopes, Malthus conceded that humans possess a high level of technological ingenuity. But he did not observe any machinery that indicated technology would soon, or ever, free humanity from toil, misery, and population checks. In fact, the effects of machines were unknown. He wrote, "Persons almost entirely unacquainted with the powers of a machine cannot be expected to guess at its effects. I am far from saying, that we are at present by any means fully acquainted with the powers of the human mind; but we certainly know more of this instrument than was known four thousand years ago; and therefore, though not to be called competent judges, we are certainly much better able than savages to say what is, or is not, within its grasp."[128] Observation of industrial capitalism's machinery granted Malthus and his contemporaries a certain level of insight into the "powers of a machine" and its "effects." However, his judgment of machines was conflicted. The "great machine" represented the ideal form of governing power. But the destructive effects of literal machines were blatant and far from ideal.

Watches and telescopes were metonymically complicit in the destructive and preservative powers of the entire economic system that would, more and more, use machines to make machines.[129] In comparison to the "great machine," the actual mechanical devices in factories were, for workers, not-so-great machines. Malthus did not need to "guess at their effects" on the working poor, because they were obvious. Malthus bemoaned the destructive elements of British industrialization and noted that capitalism's "great towns and manufactories" were home to "unwholesome occupations," "severe labour," and inadequate housing, all of which demonstrated the deleterious "powers of machines" and the failure of the "great machine" to alleviate them.[130] Factories and new machinery threw already-poor people into worse destitution, ruined their health, and otherwise brought on their early demise by cultivating drunken debauchery in manufacturing centers. The cotton industry's new spinning and ginning machinery thus exemplified how new technologies born of industrialization introduced simultaneous state wealth production and human destruction. They increased the possibility of consumption—economic and medical—and empowered employers to replace adult workers with children to run the machines.[131] Despite bearing witness to these "powers of the machine," Malthus remained committed to the belief that manufacturing, capitalism, the government, and technology could be reformed to alleviate the misery of the working and unemployed poor. He was not an "enemy to machinery," Malthus declared, and his hope in industrialization grew with each edition of the *Essay*.[132]

Thus what makes Malthus's usage of machine *antistasis* remarkable for weapons rhetoric is the way it linked the preservative and destructive aspects of technology in one banal word. I. A. Richards and Ron Greene, decades apart from each other, criticized similar machine metaphors, arguing that they cause misunderstanding and oversimplification.[133] But in Malthus's population theory, the machine metaphor's elucidation of Technē's Paradox shows that mechanistic metaphors can function in a complex way. In Malthus's case, his machine *antistasis* expressed the material and ideological functions of technology in one word, which thereby expressed the fundamental paradox of technology. Far from simplifying meaning, Malthus's machine *antistasis* laid bare that new technologies contributed to the preservative powers of industrial capitalism, while individual machines ruined lives, families, and cities and fomented class violence. Likewise, the word "technology" has come to represent both physical and conceptual danger[134] as well as the potential for safety, as did the presence of new killing technologies. His use of *antistasis* might have been casual, but the instability of meaning that the word "machine" created in the *Essay* showed both the creative and destructive characteristics of technology. Therefore, Malthus's use of the word "machine" reveals another way that his work functions as an important pre-text for understanding the development of weapons rhetoric. The metaphorical and literal switching between different meanings of "machine" granted insight into the conundrum of whether the biggest and smallest machines were advantageous or disadvantageous. They were both at the same time, and the instability of his usage of "machine" framed individual mechanical artifacts as possessing an ominous power over people. And with the further popularization and circulation of the machine metaphor, it became a commonplace that one machine could both preserve and annihilate rival populations, if not all humanity.

3. Conclusion

So if "it must be acknowledged that bad theories are very bad things, and the authors of them useless and sometimes pernicious members of society,"[135] then judging Malthus is rather difficult. Because Malthus named overpopulation as a primary technological threat and preservative force, he can be faulted for empowering Malthusians with a justification for annihilating others while at the same time commended for warning us and bringing the predicament to everyone's attention. As a pre-text for understanding subsequent weapons rhetoric, Malthus showed us that Technē's Paradox is a moral and psychological dilemma as much as it is a technological and political one. In the moral iteration, "it seems highly probable that moral evil is absolutely necessary to the production of moral excellence."[136] And in the psychological iteration, Malthus conceded that the *Essay* cultivated a gloomy outlook for humanity, but he was "subservient to the important end, of bringing a subject so nearly connected with the happiness of society into more general notice."[137] Regardless of whether Malthus is judged in retrospect as pernicious or laudable, correct or incorrect, or his principle judged as good or evil, in order to promote happiness, society has had to reflect on gloomy and unpalatable plans aimed at controlling the balance between procreation and agriculture.

Just as Malthus has an ongoing history, so, too, does the rhetoric that makes Malthus's *Essay* a compelling pre-text for understanding Technē's Paradox. The rhetorical tactics Malthus used to negotiate the Paradox—statistical sublimity, amplified generalizations, and machine *antistasis*—integrated the increasing magnitude of technological activity with a complementary rhetorical magnitude. These magnitudes can help people judge the simultaneous appearance of weapons and words about weapons. As Thomas Farrell wrote of magnitude, "in between the microbe and the gargantuan are the many truths of human scale, density, amplitude, weight, and proportion for poetic and rhetoric to invent, recover, depict, and judge."[138] Malthusian magnitudes and Malthus's arguments updated the language of weapons, war, and politics to accommodate increasing populations, technological change, and bloodshed, while the collective weight of so many demographic bodies continued to press upon the use of technologies to preserve and annihilate. This rhetorical pre-text is not a rigid frame that determines how weapons rhetors must speak and write about weapons and other technologies. Rather, Malthus circulated a few rhetorical tactics and technological concepts that later rhetors could seize for their own purposes.

Once Malthus made the willful, destructive abuses of technologies and their effects crucial variables for thinking about the problems of population management, his rhetorical tactics have been used to justify Malthusianism in all of its varieties. The Malthusian League's rationalization of birth control technologies as a way to manage overpopulation and resistance to antiabortion movements are overt examples.[139] Elsewhere, eugenics, forced sterilization, concentration camps, and genocide drew upon Malthusianism for legitimation.[140] In these ways Malthusianism has borne a technological legacy as Malthusians, neo-Malthusians, and their critics took the latent technological aspects of the "principle of population" and made them primary foci of population control.[141] After Malthus empowered them with a forceful ideology, political actors, soldiers, engineers, and scientists could use the anxiety that derived from the possible implications of overpopulation to justify extermination for the sake of saving one's own multitude. Hence, the ism that bears his name—Malthusianism—still "colonizes our future."[142] The following chapters show how statistical manipulations, hyperbolic generalizations, and polysemy remained crucial elements of Malthus's pre-textual rhetorical legacy for weapons rhetoric as new weapons began to catch up to and bring presence to Malthusian themes of overpopulation, state preservation, and anxiety about global destruction and preservation.

2

Preaching Dynamite | August Spies at the Haymarket Trial

I am an incendiary—let it be so!
—August Spies (1886)

We expect to bring into court dynamite bombs by the dozen, and until the dozens run up into barrels.
—George C. Ingham (1886)

In Chicago on May 3, 1886, the third day of a national strike for the eight-hour workday, commotion at a farm-machinery factory, McCormick's Reaper Works, interrupted August Spies's speech to the nearby Lumber Shovers' Union.[1] Spies, one of the primary labor agitators in Chicago—a savvy rhetor who had lectured around the Midwest and East Coast, and a soon-to-be martyr for the labor movement—was at the center of the forty-thousand-person strike.[2] His newspaper, *Die Chicagoer Arbeiter-Zeitung*, chronicled police abuses of labor, denounced capitalism, urged workers' solidarity, and called for revolution. It was the city's most popular German-language newspaper and the primary organ of the Chicago workers' movement.[3] Despite Spies's prominence as an agitator, there was so much commotion at McCormick's that hundreds of workers abandoned his speech and headed to the factory. Spies followed. He arrived to witness police firing at random into the crowd as strikers, police, and other combatants ducked and darted around idle freight train cars. Several strikers died, and an unknown number were injured. Spies called for retribution in an article—"Blood!"—published in the next day's *Arbeiter-Zeitung*. "With good weapons and one single dynamite bomb, not one of the murderers [the police] would have escaped his well-deserved fate," he wrote.[4] Little did he know that the eight-hour-workday strike would culminate that evening with, in fact, one single dynamite bomb that exploded at a protest at Haymarket Square, an explosion that triggered a crackdown on the workers' movement, a controversial trial, and his execution.

Spies was a skilled agitator-orator, but my focus in this chapter is as much on the crucial object of the Haymarket events—dynamite—as it is on Spies's words. I argue that the presence of dynamite interacted with polysemy and paradox when Spies preached dynamite. In turn, Spies delved into and exploited the rhetorical instability of weapons advocacy when, in close proximity to dynamite, he negotiated Technē's Paradox by strategically using language's plasticity. By importing the contingencies of linguistic meaning and rhetorical flux into his antistate argumentation, it was as if Spies had taken Friedrich Nietzsche's concept of the "mobile army of metaphors, metonyms, and anthropomorphisms" and imagined it as a tactic for dealing with the presence of dynamite and its contingent, unstable meanings.[5] Yet despite how much Spies spoke of dynamite, dynamite's material presence also motivated action at Haymarket.

Dynamite and oratory mixed on the night of the Haymarket bombing. A group of mainstream press reporters joked with the Parsons family just before gathering at Haymarket. "Have you any dynamite about you?" the *Times*'s Edgar Owen asked Albert Parsons, Spies's colleague, the husband of labor activist Lucy Parsons, and the editor of an anarchist newspaper called *The Alarm*.[6] Having had difficulty rounding up speakers for the event, Spies started the Haymarket rally late with a rather innocuous twenty-minute speech. Parsons followed. He exhorted, "in the interests of your love of liberty and independence, to arm, to arm yourselves!"[7] But Parsons had just demurred, saying that "I am not here for the purpose of inciting anybody, but to tell the truth."[8] The meeting was peaceful, as even attendee Chicago mayor Carter H. Harrison attested. Samuel Fielden spoke last, but he did not speak long. With a storm blowing in, the crowd dwindling, and the rally nearing its conclusion, a force of 176 police officers under the direction of Captains John Bonfield and William Ward closed in on the speakers' wagon in double time. Ward ordered the crowd to disperse. Just as Fielden objected that the event was "peaceable," a dynamite bomb sailed into police ranks and exploded, killing at least one policeman, Mathias Degan. Chaos erupted as the police began firing their revolvers into the crowd "in wild confusion" and clubbing protestors as they fled.[9] According to some reports, Lieutenant James Bowler yelled, "Fire and kill all you can!"[10] Bonfield secured a gun from the hand of a dead officer and blazed away with two weapons.[11] During the mayhem, an apparent assassin planted in the crowd attempted to shoot Spies in the back. Spies's brother, Henry, intervened to save Spies and got shot in the groin. Seven policemen died, most from friendly fire, and dozens were injured. An unknown number of civilians died and were injured.

In the days after the Haymarket bombing, workers' proximity to dynamite and political allegiances determined accountability for the attack. Chicago police rounded up scores of agitator-anarchists and ended up charging Spies, Parsons, Fielden, Adolph Fischer, George Engel, Michael Schwab, Louis Lingg, and Oscar Neebe with conspiring to throw the Haymarket bomb. Although Lingg had spent much of his short time in America manufacturing dynamite bombs, including those of the "czar bomb" variety that exploded at Haymarket, and Neebe was a luckless supporter of the workers' movement, the charges for the other six men derived from their publicity work—orating, writing, printing, and publishing "incendiary" anarchist rhetoric that touted dynamite as the means to liberate workers from religious, economic, and political oppression.[12] Spies had handled, written, and spoken of dynamite so often that he could not free himself of the conspiracy charge. He had lived by calling workers to arms. And the Chicago legal system would execute him for it, but not before he preached dynamite one last time in the courtroom address he delivered upon receiving the death sentence.

In his "Address to the Court," Spies appropriated the phrase "to preach dynamite" from mainstream journalists to describe his oratory.[13] Preaching dynamite was an apt phrase for encompassing not only the numerous appeals to destroy capitalism, the government, their edifices, and their advocates with dynamite, but also the unqualified high praise of and faith in dynamite that circulated in Chicago and around the world circa 1886.[14] By entering into this commonplace dynamite discourse, Spies entered into contact with Technē's Paradox, and preaching dynamite was a paradoxical rhetoric of technology, much like Malthus's, that summoned technology in order to combat a world that was more and more beholden to it. But Spies was not a would-be social engineer who sought to maintain an aging republican society amid the industrial revolution. He was an activist seeking to deploy revolutionary technologies to overthrow capitalism. Spies and his coanarchists had developed a technological rhetoric of terrorism to advocate a classless technological society.

This chapter's middle-way approach contends that both Spies and dynamite shaped the character of preaching dynamite at the Haymarket events. Of course, Spies spoke about much more than dynamite in his two-hour courtroom address:

he argued that he had nothing to do with the bombing; that because the bomber remained unknown, the charge of conspiracy was not provable; that Chicago's corrupt police had kidnapped his main alibi; that the state had bought the witnesses who testified against him and had handpicked a jury predetermined to convict; that the prosecution had tried him for believing in anarchism rather than murder; and that the state would murder innocent men.[15] But his oratory was also the result of the material interaction of dynamite with Spies. At Haymarket, a particular weapon influenced humans, and humans used a particular weapon to influence. This chapter seeks the middle-way path by zeroing in on and bracketing dynamite—the word and the weapon—to explain how Spies negotiated Technē's Paradox at the Haymarket trial. For the purposes of the following analysis, I therefore focus on the particular object of dynamite, its signifier, and the arguments it warranted.

Before I turn to the address itself, in the first section I examine how both Spies and the technology of dynamite partly controlled how the trial unfolded. In section 2, I analyze Spies's "Address to the Court" to demonstrate two prevalent rhetorical tactics occasioned by preaching dynamite. First, Spies delved into the polysemy of the term "dynamite" by drawing the court's attention to the word's multiple meanings. Second, the presence of Technē's Paradox empowered Spies to construct arguments that reversed the accusation of conspiring to commit violence back onto the state. Following this, I examine the post-Haymarket period in order to demonstrate how preaching dynamite and dynamite itself limited the capacity of the Chicago anarchists to motivate solidarity and revolution. The chapter concludes with some remarks about the materiality of preaching dynamite.

1. Dynamite at the Haymarket Trial

By 1886, the dynamite bomb had worked its way into oratory, journalism, and political philosophy, and anarchists had come to uphold dynamite as *the* singular technology that would sweep capitalists from power and destroy their governments.[16] And dynamite itself seemed to compel the course of the Haymarket trial when the substance worked as evidence for the prosecution, corroborated Spies's calls to arms, refashioned Spies's loose allegiance to anarchism and dynamite into an unbreakable bond, and intimidated the courtroom audience, all of which empowered the court to convict Spies of conspiracy to commit murder.

The fate of Chicago and the fate of Spies seemed in 1886 to depend upon the presence of dynamite. Overall, the inseparability of Spies's rhetoric and dynamite demonstrates that, in terms of Technē's Paradox, the persuasive force of a weapon's physical presence to preserve or destroy is as important as language about the weapon.

The hodgepodge of dynamite artifacts used as evidence by the prosecution proved sufficient for the court to find Spies guilty of conspiring to commit murder. His proximity to the weapon in the courtroom accomplished what the prosecution could not accomplish via direct evidence, argumentation, testimony, and cross-examination. In his opening remarks, prosecuting attorney George C. Ingham said, "We expect to bring into court dynamite bombs by the dozen, and until the dozens run up into barrels. No bomb which we shall trace to these defendants can have any possible legitimate purpose. We shall show by men of science that dynamite bombs cannot be used for anything else but for cowardly and atrocious murder."[17] During police captain Michael Schaack's testimony and as the prosecution wrapped up its case, a wide sampling of evidence was "strewed" and "piled" on tables, on the floor, and "everywhere around" the courtroom: bomb fragments removed from policeman Mathias Degan's corpse, bloody, shrapnel-pierced police uniforms of dubious origin, fragments of objects blown up by police with dynamite seized during illegal raids of suspected anarchists, photographs of random bombs, Engel's unused metal furnace, bomb parts that Spies kept in his desk and used to threaten a mainstream journalist, and dynamite found, perhaps, in a closet of the *Arbeiter-Zeitung* building.[18] According to the *Tribune*, "The centre of the court . . . seemed a cross between a dynamite arsenal and a newspaper office."[19] Injured policemen displayed and described their injuries. The prosecution could not prove the veracity of most of its evidence or that most of its evidence was directly connected to the Haymarket bombing, much less that any of it proved the defendants guilty of conspiracy.[20] Yet these eloquent objects materialized Spies's preaching of dynamite as if they had become capable of communication.[21] In the words of Perelman and Olbrechts-Tyteca, the "real things," the artifacts, were "expected to induce an adherence that [their] mere description would be unable to secure."[22] The dynamite artifacts acted upon the jury's emotions, substantiating the prosecution's weak case.[23] Thus, as the witnesses and dynamite artifacts mingled and interacted upon each other within the courtroom, the mere presence of dynamite framed Spies as a bomb-thrower's accomplice.

The presence of dynamite gave force to the anarchist documents that were strewn amongst the dynamite artifacts, despite the prosecution's failure to show, aside from their proximity to Spies in the courtroom, that the evidence written by Spies and other "dynamite orators" was linked to the Haymarket bomb.[24] Spies had spent much of his prolific career writing, orating, and publishing documents that reiterated militant and commonplace anarchist, communist, and socialist slogans. The infamous "Revenge Circular" that Spies wrote while incensed at the bloodshed of the McCormick's riot well exemplifies his typical call to arms. "REVENGE! Workingmen, to Arms!" the circular read in bold letters. The simplicity of this slogan both obscured the economic impetus of the eight-hour-workday strike and ideographically associated liberty with armaments as a "material idea."[25] To paraphrase Mark Moore's assessment of how material ideas function in handgun rhetoric, these calls to arms in the name of workers' freedom were "an ideograph in synecdochic form," such that dynamite came to "represent an ultimate social conflict (life versus liberty)."[26] The "Revenge Circular" concluded in language that could have emanated from a number of revolutionary nineteenth-century political philosophies: "If you are men, if you are the sons of your grand sires, who have shed their blood to free you, then you will rise in your might, Hercules, and destroy the hideous monster that seeks to destroy you. To arms we call you, to arms!"[27] At the trial, Spies called his flyer's response to the McCormick's riot "an extravagance," but it was not an anomalous document.[28] Spies's speech to the Lumber Shovers' Union, his article that foreshadowed the Haymarket bombing, and his "Revenge Circular" were workaday affairs, little different from his full-time agitation for workers' rights. Yet the banality and vagueness of these documents did not matter for the prosecutors who presented Spies's "Revenge Circular," "Blood!," and dozens more "inflammatory" slogan-laced *Arbeiter-Zeitung* articles as evidence. For the prosecution, the ideographic anarchist calls to arms that circulated throughout the city were as "empirically 'present'" to Chicagoans as the Haymarket bomb.[29] Moreover, the prosecution evidenced Spies's journalism by reproducing it on plain sheets of paper without the surrounding journalistic context, which removed the background of worker oppression and labor rights that was evident in the excised portions of these articles, newspapers, and speeches. Stripped of any industrial or governmental accountability, the copresence of weapons and slogans that was meant to build solidarity in revolution confirmed the proximity of Spies and his colleagues to the murder weapon and verified that they were guilty of being anarchists. Spies accepted the weaponization of his rhetorical persona. "I am an incendiary—let it be so!" he later declared in the autobiography he wrote in prison.[30] The Haymarket dynamite bomb thus proved persuasive of Spies's guilt in a context in which he called workers to arms again and again.

When Spies, dynamite, language, and anarchism became inseparable at the trial through material presence and repeated association, these elements together formed the Spies "orator-machine" assemblage. According to Matthew S. May's analysis of labor agitator "Big Bill" Haywood, "the function of the orator-machine is to activate and reproduce the flows of resistance that . . . represent the potential to transform the random noise of the fits and starts of the interconnected organs of capitalist machinery into a permanent and irreversible distortion of the system itself."[31] Spies was such an orator-machine whose "Chicago idea" of revolutionary unionism used "different compositions of living organs and machines," such as workers, owners, politicians, factory machinery, printing presses, and dynamite to demolish capitalism.[32] The "Chicago idea" held that the nascent militant labor unions in the city were forming the autonomous collectives that would organize the postcapitalist commonwealth.[33] When the state tried Spies for "endeavoring to make anarchy the rule," as lead prosecutor Julius S. Grinnell put it in his opening statement, the court cared little for the nuances of anarchist theory.[34] Instead of his advocacy of revolutionary unionism, it was his possession of dynamite and his desire to destroy the government that were the anarchist traits that Spies had to possess in order for the charge of murderous conspiracy to stick. Have them he did. Thus, the Spies "orator-machine" was an assemblage that united rhetor, rhetoric, weapon, and political philosophy such that Spies, dynamite, and anarchism all came to represent each other metonymically by being embodied in the same person.

It might have surprised Spies, his codefendants, and the court to know that the Chicago anarchists are credited with creating a school of anarchism, for the state considered anarchism tantamount to unprincipled annihilation, and the codefendants considered anarchism one of many possible modes of agitation. The Chicago labor movement's initial embrace of anarchism derived more from the provocative force afforded by appropriating the divisive terminology with which Chicago's authorities had labeled it than from a strict, theoretical allegiance to the political philosophy. According to Parsons, journalists and politicians began calling Chicago labor agitators "anarchist" in 1881 in order to portray workers in terms meant to be as frightening and odious as possible to Chicagoans. "We began to allude to ourselves as anarchists, and that name which was at first imputed to us as a dishonor, we came to cherish and to defend with pride," Parsons wrote in his autobiography.[35] The agitators therefore appropriated the term "anarchist" to make their dissent appear as aggressive as possible, and appropriating the anarchist label entailed embracing dynamite. All institutions that propped up capitalism were therefore open to the threat of being targeted with anarchist dynamite, so the defendants' rhetoric vacillated between attacks on religion, federal and local governments, capitalism, and nonrevolutionary labor organizations in Chicago, depending on the different contexts and audiences faced by "orator-machines" like Spies and Parsons. The prosecution made no effort to understand what Spies advocated, and thus the court portrayed him as a simple amalgamation of dynamite, inflammatory rhetoric, and anarchism rather than as a conceptually complex and technologically vibrant political orator.

Dynamite also made its presence felt at the conclusion of the Haymarket trial on October 7, 1886. Judge Joseph E. Gary denied the defense's request for a retrial and upheld the court's death sentences for Spies, Parsons, Fielden, Fischer, Engel, Lingg, and Schwab. Oscar Neebe received a fifteen-year prison sentence. Judge Gary, however, did grant the eight men a chance to address the court. Spies delivered the first speech, and the audience was attuned to how he would speak about dynamite. A widely circulated rumor held that anarchists had planted a massive number of dynamite bombs under the courthouse and planned to blow up the building if the court delivered a guilty verdict. There were no such bombs. Yet the dynamite at the Haymarket trial exhibited the force, energy, and intensity that Jane Bennett has suggested one should expect to discover if one is cognizant of how technology and people interact. In 1886 dynamite was "vibrant" in its power "to exceed its status as an object" as it "manifested traces of independence or aliveness" that were outside of human volition and that produced "effects in human and other bodies."[36] Dynamite was so vibrant at the Haymarket trial that it made its presence felt, even when absent, by provoking anxiety and rumor.

Thus, dynamite infiltrated the trial as emotion-stirring evidence, infiltrated Spies's calls to arms, partly defined the concept of anarchism that made Spies an anarchist, and struck fear into the courtroom audience. At the end of the trial, after a brief deliberation, the jury found the prosecution's case sufficient to convict the defendants of conspiracy. Thus, in 1886, dynamite was at the epicenter of what Thomas Farrell has called rhetorical *dynamis*, or "power in its nascent state."[37] This nascent power composed of the mistrusted political philosophy of anarchism, the vibrant materiality of dynamite bombs, and dynamite advocacy constructed the meaning of political contestation in Chicago as Spies drafted his final speech. And, as he negotiated Technē's Paradox in his "Address to the Court," Spies had not only to counter the independent persuasive force possessed by dynamite, but also to capture it in order to mount his defense.

2. Spies's Address: Strategic Polysemy and Paradoxical Turnaround Arguments

One might think that, given the gravity of the situation, with his life hanging in the balance and the audience worried about being dynamited, Spies might have refrained from preaching dynamite in his courtroom address, perhaps to seek a less inflammatory vein of argument in order to set the tone for his later appeal. He did not. Spies utilized the rhetorical instability afforded by preaching dynamite and negotiated Technē's Paradox by fostering "competing modes of understanding" a violent event, to use the words of Stephen Browne's analysis of rebel slave Nat Turner.[38] First, Spies used strategic polysemy to construct a meaning of dynamite that did not include an exhortation to hurl a bomb at Haymarket, and second, he hurled the accusation of conspiracy back upon the state with turnaround arguments warranted by the presence of dynamite and hinging upon dynamite's simultaneous destructive and preservative powers. For the purposes of the following analysis, though, I will diverge from traditional rhetorical criticism in my unwavering concentration on a particular object (dynamite), its signifier, and the arguments it warranted. The scope of this analysis is hence not a broad look at the entirety of Spies's speech, which one might expect from public-address scholarship. Rather, I aim to tread the middle way between the material and rhetorical turns by doing some nontraditional object-oriented bracketing within a traditional framework of rhetorical criticism.

Spies established strategic, polysemous meanings of dynamite by speaking of the explosive for multiple rhetorical purposes. Leah Ceccarelli defined a polysemous text as one that is "open to a certain number of different readings by different groups and classes of viewers," such that the different readings "indicate a bounded multiplicity, a circumscribed opening of the text in which we acknowledge diverse but finite meanings."[39] The more specific concept of strategic polysemy is a type of rhetorical invention whereby a "calculating rhetor" attempts "to bring different audiences, through different paths, to a point of convergence in the acceptance of a text."[40] In his defense, Spies used strategic polysemy by speaking of dynamite as a metaphor for revolution—a metaphor that he contextualized with other violent metaphors in contemporaneous journalism—by speaking of dynamite as a means of self-defense, and by speaking of dynamite as an unremarkable facet of the historical progression toward capitalism's demise. Spies's strategic polysemy might make him appear ambivalent toward the weapon, but as Sandra Harding has noted, technologies in general "have 'interpretive flexibility' such that their expressive meanings are context-dependent."[41] Each meaning of "dynamite" was crucial to each usage as Spies worked to show that his invocations of dynamite were aimed at preserving the working population, not at inciting them to throw the Haymarket bomb.

Before examining how Spies used strategic polysemy, it is important to examine why the state and other authorities interpreted his dynamite advocacy as possessing a singular, nonpolysemous, literal meaning that Spies was directing his audience to start lighting fuses and throwing dynamite bombs. At a time when anarchists, capitalists, journalists, and soldiers could use dynamite simultaneously for contested military, political, industrial, and symbolic purposes, Judge Gary's court insisted that Spies meant exactly what he said when he called for violence. I. A. Richards would have said that the authorities suffered from the "Proper Meaning Superstition," which holds that the meaning of words is always stable and never context dependent.[42] This insistence that Spies had called for the Haymarket bombing started as soon as Chicago's mainstream newspapers fired up their printing presses to report the incident, when farm machinery magnate Cyrus H. McCormick Jr. and the *Chicago Tribune* immediately blamed Spies and his "incendiary" language for the bombing.[43] At the trial, the court concurred that the bomb had materialized Spies's calls to arms, and throughout the trial, the court combined Spies's use of the word "dynamite" and the Haymarket bomb as evidence of his conspiratorial guilt, even though Spies repeatedly said in the days leading up to the Haymarket bombing and on the Haymarket speakers' wagon that it was not yet time for antistate violence. At long last, the authorities asserted, an anarchist had thrown a bomb at Spies's behest. The prosecution's case therefore concentrated on the connection between Spies's rhetoric and the unknown bomber's weapon. Spies had pushed his exhortations to dynamite the capitalist establishment to the limit, so that when the Haymarket bomb exploded, his oratory became coterminous with bomb throwing. By establishing that Spies's rhetoric was as dangerous as dynamite, the causal link between Spies and the Haymarket bombing rendered the various other meanings inherent to Spies's dynamite oratory inconsequential to the verdict. Of course, Spies was not in full command of his strategic polysemy since he could not control how the state, the jury, the press, and the public interpreted his words. Yet the debate about whether to understand Spies's calls to arms as literal indicated that, regardless of intent, when he spoke of dynamite and other weapons, the presence of polysemy meant that his volatile and unstable words were open to a wide range of interpretations, especially surrounding the tumultuousness of a volatile and unstable event like the police advance on the Haymarket rally.[44]

Confronted by the prosecution's interpretation of his calls to arms, Spies attempted to establish the multiple meanings of the word "dynamite" by first summoning the weapon as an abstract metaphor for revolution. Speaking in what fellow anarchist agitator and trial observer Albert Currlin called "a clear, steady voice," both "brilliant and sarcastic," Spies asserted that "from [the state's] testimony one is forced to conclude that we had, in our speeches and publications, preached nothing else but destruction and dynamite."[45] Spies embraced the accusation. "We have preached dynamite!" he proclaimed. "We have said to the toilers, that science had penetrated the mystery of nature—that from Jove's head has sprung a Minerva—dynamite!"[46] Dynamite represented the pinnacle of technological and scientific innovation for anarchists, and although the anarchists' calls to arms materialized at Haymarket, as much to their surprise as anyone else's, the tossing of a dynamite bomb might have been expected. Since Alfred Nobel had invented the explosive in 1867 by combining nitroglycerine and kieselguhr into a putty, wrapping it in cardboard, and adding a blasting cap, dynamite had become freighted with an increasing magnitude of metaphorical power for a couple of reasons.[47] First, poor workers could produce large amounts of dynamite in their basements—unlike other popular weapons of the time, such as Winchester rifles, Colt revolvers, and Gatling guns, which workers had more difficulty obtaining. With dynamite costing about forty cents per pound and individual bombs costing between six and fifteen cents, almost any daring individual could obtain or make them.[48] Second, as propounded by Johann Most's anarchist treatise *Revolutionäre Kriegswissenschaft* (1885), the successful dynamite assassination of Russia's Czar Alexander II in 1881 demonstrated that the weapon was a viable means to commit "propaganda by the deed."[49] In the hands of all workers, dynamite appeared to match the destructive power of state militaries. As such, dynamite gained prominence in anarchist political philosophy as the antidote to capitalism.[50] An invocation of dynamite was thus for Spies and the Chicago anarchists a metaphor for the death of capitalism.

Contextualizing his invocations of dynamite with other violent metaphors empowered Spies to mitigate the violent connotations of dynamite by drawing attention to the rhetorical instability of his violent words and decrying the hypocrisy of the charges against him. Spies quoted a Wisconsin newspaper, the Fond du Lac *Commonwealth*, which demanded, "To arms, Republicans!"[51] The paper appeared to incite Republicans to violence. "Every Republican in Wisconsin should go armed to the polls next election day. The grain stacks, houses and barns of active Democrats should be burned[,] their children burned and their wives outraged," it read. "Shoot every one of these base cowards and agitators."[52] Spies expressed irritation that his metaphors led to a death sentence when this list of atrocities went unnoticed, much less unpunished. If the court considered the *Commonwealth*'s exhortation to murder political opponents' children and rape their wives a metaphor for voting, then Spies sought an explanation for why his metaphors should be interpreted literally as advocating throwing bombs at Haymarket. "How does the *Arbeiter-Zeitung* compare with this?" he demanded.[53] Incredulous of the court's hypocrisy, Spies insisted, "Show me a daily paper in this city that has not published similar articles!"[54] Spies had a point. In February 1885, the *Tribune*, for instance, detailed how to make what we now call time bombs, as well as hat bombs, bottle bombs, fuseless grenades, various types of gunpowder, and dynamite bombs.[55] His dynamite oratory was no more incendiary than and just as banal as typical political discourse, but the *Commonwealth*'s example indicated how the polysemy of dynamite became mobilized for contested political purposes. The government, Spies suggested, seemed to authorize an interpretation of the *Commonwealth*'s exhortations to commit atrocities as metaphors, but not Spies's calls to arms. Thus an injustice, as a literal call to arms was clearly just one of several possible interpretations of dynamite.

Invoking the use of dynamite for self-defense was the second aspect of Spies's strategic polysemy. Instead of framing the Haymarket bombing as retaliatory or offensive violence, he represented it as an act of preservation by the unknown bomber in the face of the police regiment's rapid advance. To emphasize the defensive character of the Haymarket bombing, Spies juxtaposed the official silence that accompanied state-sanctioned attacks on workers with the "terrific howl" that arose after Haymarket, "when, on one occasion, a workingman in self-defense resisted the murderous attempt of the police and threw a bomb and for once blood flowed on the other side."[56] The police at Haymarket had not yet committed any violence that night, but Spies insinuated that their past propensity to club and shoot protestors gave the bomber a just impetus to hurl his infernal machine. Spies referenced many such incidents of antilabor violence, such as when railroad magnate Jay Gould's "hirelings" (Pinkerton detectives) killed six workers in East St. Louis during the Great Southwest Railroad Strike of 1886.[57] The commonness of antiworker violence thus exemplified that in 1886, dynamite meant self-defense for Chicago's workers.

The third element of Spies's strategic dynamite polysemy entailed framing the weapon so that it could be interpreted as a sign of the unstoppable movement of history. He said, "I may have told that individual who appeared here as a witness that the workingmen should procure arms, as force would in all probability be the *ultima ratio*; and that in Chicago there were so and so many armed, but I certainly did not say that we proposed to 'inaugurate the social revolution.'"[58] Spies did call for revolution, but not for immediate revolution. The bomber, Spies suggested, went astray, misunderstanding the timing and purpose of his calls to arms. Without massive, widespread support throughout Chicago (to form the heralded Chicago Commune), or the country (for a general strike), the state would crush any attempted revolution, so the bomb's effects were historically negligible. In his autobiography, he reflected on the absurdity of the charge that he conspired to overthrow the US government with "*one* bomb to be thrown 'at some intersection of a street!'"[59] Spies's demand that workers must use dynamite would only transform from meaning self-defense into an exhortation to commit immediate, offensive violence when workers had attained the support and means to attain victory, and not until then. With the futurity of the revolution established, Spies minimized his invocations of dynamite as insignificant instants in a long, inevitable historical progression.[60] Dynamite happens. Thus, the state should not blame him if the bomber misunderstood the word that Spies used as a banal metaphor for revolution, a call for self-defense, and a general proclamation about the eventual demise of capitalist governments.

As Spies invoked multiple meanings of dynamite, he indicated that he was both accountable and not accountable for the Haymarket bombing. On one hand, he was accountable for Haymarket by being an agitator who encouraged violence as a general tendency in revolutions, even though he discouraged it in Chicago. On the other hand, he argued that he was not accountable because he was a mere minor agent in a larger historical movement. According to this historicism, the state should not hold him accountable for heralding what history dictated was going to happen regardless of whether Spies had ever even existed. In the grand historical scheme, he argued, his violent speech was as "whimsical" as he claimed his adversaries' authority was.[61] And therefore, his violent language was meant to be understood as an exhortation to commit murder no more than anyone else's. Thus, Spies identified and exploited the polysemy of dynamite by at once claiming it as mere rhetorical whimsy and as the central symbol of anarchist violence and as the means of self-defense.

Another curious possibility afforded by paradox in general, and the rhetorical instability of preaching dynamite in particular, is that of *antistrephon*, or turning arguments against each other.[62] Turnaround arguments appropriate the opposition's language, arguments, and evidence to argue one's own position, a tactic well suited to paradoxes that hold two seemingly opposed positions at once. By using turnaround arguments, rhetors may do more than negotiate paradox. They may exploit it, and Spies did exactly that. When the presence of dynamite at the Haymarket events pushed Spies into a direct confrontation with the weapon, he used the rhetorical instability afforded by Technē's Paradox to construct turnaround arguments about how dynamite facilitated simultaneous political destruction and preservation. The Paradox situated the legal relationship of revolutionaries to the government. What the state viewed as preservative—capitalism—Spies viewed as destructive, and what the state viewed as destructive—anarchism—Spies viewed as preservative. Spies therefore turned the state's evidence against him into evidence against the state. As Spies spun defense from attack and spun attack from defense, the concepts of destruction and preservation swirled around each other with increasing intensity. Destruction, according to Spies, meant capitalism's annihilation of workers, while destruction, according to the state, meant anarchism's annihilation of democratic society. Preservation, according to Spies, meant preserving the masses by annihilating the forces of capitalism, while preservation, for the state, meant preserving the masses by annihilating the labor movement and anarchism.

Spies first indicated that he would turn the state's case into his defense, and turn his defense into an attack on the state, by quoting Marino Faliero, a fourteenth-century doge of Venice. "*My defense is your accusation; the causes of my alleged crime your history!*" he said.[63] For Spies, the state was guilty of provoking the Haymarket bombing because of its violent maintenance of capitalism, and therefore he used Faliero's age-old commonplace to charge the police, the government, and the court—the "real revolutionists . . . conspirators and destructionists"—with the "heinous conspiracy to commit murder."[64] Dynamite served as the material proof of this turnaround. The fact of the Haymarket bombing demonstrated that this oppressive history endured, and that the prosecution only needed this one bomb to put an end to the eight-hour-workday strike with police brutality, illegal arrests, a trial, and executions. And although nobody could prove Spies threw the bomb, Spies could prove the bombing was justified by so much injustice. Spies thus expressed that the weapon will facilitate the ultimate destruction or preservation of society and civilization depending on which political system ends up victorious.

As his address unfolded, Spies turned one of the state's accusations about his involvement with an alleged dynamite conspiracy into an accusation that the state had conspired to commit violence against impoverished, striking coal miners in Ohio. In the trial, the prosecution had presented the 1884 strike against the Hocking Valley Coal and Iron Company as evidence of the barbarity of the labor movement in general and the barbarity of Spies in particular, contending that he had conspired with Johann Most to transport dynamite to the strikers. Spies, however, used the incident as proof of capitalism's barbarity. He paraphrased the prosecution, agreeing that "I might have destroyed thousands of innocent lives in the Hocking Valley with that dynamite."[65] But whether the event represented capitalist or anarchist destruction was up for grabs amid the paradoxical instability of preaching dynamite. Spies seized upon the fact of the strike and used it as his own evidence against the state to argue that capitalism's crimes were greater than his: "I saw hundreds of lives in the process of slow destruction, gradual destruction. There was no dynamite, nor were they Anarchists who did that diabolical work. It was the work of a party of highly respectable monopolists, law-abiding citizens, if you please."[66] Rather than observing miners armed with dynamite, he had witnessed eighty Pinkerton guards armed with Winchester rifles looming over his speech to the Hocking Valley miners.[67] The danger to society posed by an idle mine could not compare to the danger to society posed by killing its productive population. Thus, Spies asserted that the Hocking Valley strike, and even the would-have-been use of dynamite by strikers, which might have "blown some of those respectable cut-throats to atoms," would have been the truly preservative actions.[68] Spies thereby exploited Technē's Paradox to show that dynamite, far from representing the destruction of society, could preserve it.

Spies used such state violence to make another rhetorical turn, again warranted by the presence of dynamite in government and anarchist arsenals, which reversed the charge that anarchists were despicable "monsters" and "fiends." Using a rather common trope, Spies transformed the monster metaphor into a description of "monstrous" capitalist society.[69] Spies called capitalism a "fiend," and this "fiend who has grasped her [society] by the throat; who sucks her life-blood, who devours her children," did not deserve preservation, but rather destruction.[70] In contrast, "Anarchism does not mean bloodshed; does not mean robbery, arson, etc.," he said, for such "monstrosities" are "characteristic features of capitalism."[71] Anarchist dynamite did not appear quite so destructive when compared to the atrocities committed by the industrial-capitalist "fiend" against innocent children. By conveying total enmity, the monster metaphor helped Spies imbue his adversaries with sinister and diabolical attributes worthy of being dynamited. Chicago police captain John Bonfield, for one, had "a visage that would have done excellent service to [Gustave] Doré in portraying Dante's fiends of hell."[72] The "monster" had no positive attributes; the "monster" had no preservative power; the "monster" intended only bloodshed; the "fiend" attacked without mercy; the "fiend" was industrial capitalism. As demonstrated by the Hocking Valley miners' strike, any pretense toward societal preservation on the part of the state was, according to Spies, a dangerous ruse. The "demons of hell" would thus admire the success of their most conspicuous compatriot—the US government.[73]

Spies expanded his critique of the state by using *antistrephon* to redefine American democratic capitalism as, first and foremost, much more destructive than anarchist dynamite bombs. What the state wanted to preserve as the basis of societal organization—industrial capitalism's monstrous effects on workers—were destructive in contrast to the "constructive science" of socialism.[74] He therefore contrasted the small magnitude of violence at Haymarket with the dreary outlook that would prevail with the preservation of the American government:

> It means the preservation of the systematic destruction of children and women in factories. It means the preservation of enforced idleness of large armies of men, and their degradation. It means the preservation of intemperance, and sexual as well as intellectual prostitution. It means the preservation of misery, want, and servility on the one hand, and the dangerous accumulation of spoils, idleness, voluptuousness and tyranny on the other. It means the preservation of vice in every form. And last but not least, it means the preservation of the class struggle, of strikes, riots and bloodshed.[75]

The preservation of "the conspirators and destructionists" visible in microcosm at the Hocking Valley mine strike was not an isolated manifestation of state power. The vast array of mechanical power—industrial and agricultural machines, railroads, telegraphs, steam power, dynamite, rifles, revolvers, Gatling guns, and gallows—that had been monopolized by corporations and politicians demonstrated to the court, Spies hoped, that capitalists and the government were more prone to destruction than anarchists.[76] For these reasons, Spies lauded the Haymarket bombing. "If I had thrown that bomb, or had caused it to be thrown, or had known of it, I would not hesitate a moment to say so. It is true that a number of lives were lost—many were wounded. But hundreds of lives were thereby saved! But for that bomb, there would have been a hundred widows and hundreds of orphans where now there are few," he proclaimed.[77] Spies disavowed any personal bloodlust, but the Haymarket bombing, although in the end futile, was an act of worker preservation aimed at arresting violent oppression.

When Judge Gary's court both scrutinized Spies's rhetoric as a primary element of its prosecution and insisted that the ubiquitous word "dynamite" represented the singular Haymarket bomb, Spies acquired a platform from which to probe the complexities and instabilities of preaching dynamite. His exploitation of the polysemy of the word "dynamite" and construction of paradoxical turnaround arguments that were warranted by the presence of dynamite revealed that preaching dynamite exposed the state's hypocritical attitude toward political violence at the same time that it failed as a legal defense. The limitation of preaching dynamite pushed back on Black Friday—execution day for August Spies, William Parsons, Adolph Fischer, and George Engel. Although Illinois governor John Peter Altgeld eventually pardoned Lingg—who had committed suicide by dynamite in prison—as well as Fielden, Schwab, Neebe, and the four executed men in June 1893, preaching dynamite brought destruction rather than liberty to the Chicago anarchists, and it corroborated Spies's assertion that the state would continue to use political violence to destroy workers rather than to preserve them.

3. The Political and Material Limits of Preaching Dynamite

For all the anarchists' calls to dynamite capitalism and to trust in dynamite as the agent of social change, the moment a dynamite bomb exploded at Haymarket, it failed to spur revolution. Despite incessant heralding, the moment for violence had not arrived. The bomb inspired neither a Chicago Commune nor a national uprising; much less did it inspire the overthrow of capitalism. So in terms of the middle way between the object and the word, what did preaching dynamite do for Spies and the Chicago anarchists? To paraphrase Charles Griffin's assessment of the United States' symbolic use of nuclear submarines in the Cold War, the Haymarket bombing's "technological spectacle . . . magnified" the "propagandistic affordances and constraints" of the weapon and its rhetoric.[78] Preaching dynamite worked against the anarchists by accentuating the material limits of advocating political violence as a practical way for agitators to empower workers in the face of overwhelming political and military opposition. In post-Haymarket Chicago, the limits of preaching dynamite were demarcated when the technical capacity of dynamite was revealed, contemporaneous labor agitators judged violent rhetoric a failure, the state turned anarchism's advocacy of dynamite against the anarchists, and Spies provided the labor movement with a symbol more enduring than dynamite by promoting his martyrdom.

The possibilities and limits of preaching dynamite on political agency proved severe, and labor advocates tended to judge the persuasive strategy a failure in the aftermath of Haymarket.[79] In one simple respect, preaching dynamite proved self-defeating because it misrepresented dynamite's technical capacity to wreak destruction, and the misrepresentation of dynamite's destructive power misrepresented the political agency of those who wielded it. Even Johann Most, one of the foremost dynamite advocates who proclaimed that "a girdle of dynamite encircles the world," admitted in *Revolutionäre Kriegswissenschaft* that when the weapon exploded, the limited amount of damage wrought exposed the commonplace amplification of dynamite's destructive power as mere hyperbole.[80] With five times more explosive power than gunpowder, dynamite was the most powerful explosive of the time, and its effects seemed catastrophic. As such, dynamite was touted as capable of blowing up everything and everyone in a single blast. *The Alarm*, for example, reprinted US Army Gen. Philip Sheridan's oft-quoted exposé of "pocket bombs" from his 1884 Annual Report. "One man armed with a dynamite bomb is equal to one regiment of militia, when it is used at the right time and place. . . . One dynamite bomb properly placed will destroy a regiment of soldiers," Sheridan claimed.[81] Dynamite may have seemed to even the power disparity between workers and the government, but it did not in fact do so. One bomb carried in a pocket did not cause such massive death and destruction even when exploded in the midst of a police regiment's tight formation at Haymarket. Dynamite did not possess an unlimited capacity to annihilate, and therefore neither did the anarchists who touted, manufactured, and used it. The technical capacity of the weapon showed that state militaries would crush bomb-throwing revolutionaries.

The threat of random dynamite bombings stoked public fear nonetheless, but the fear also proved self-defeating for anarchism.[82] Dynamite was too scary. Its destructive capacity appeared to outweigh its preservative capacity, and the weapon ended up alienating workers instead of motivating their solidarity.[83] William Holmes, a colleague of the defendants, wrote in 1892 that it was undesirable for labor advocates to "extol the excellence of dynamite as a factor in the coming crisis," arguing that "much of this kind of agitation we have had in the past. And what have been its legitimate fruits? The scorn and hatred of the very class [the working people] whom we most desire to win; the bitter enmity and persecution of the authorities; the contempt of capitalists; and the antagonism of all classes."[84] In an 1893 letter to Lucy Parsons, Chicago socialist-labor agitator George Schilling corroborated Holmes's assessment. Schilling called the Chicago anarchists' rhetorical tactics "mistaken methods" because they "lead to greater despotism" and "terrorize the public mind and threaten the stability of society with violence."[85] The unceasing threats of violence, appeals to force, and appeals to fear inherent in preaching dynamite established that such tactics of labor agitation were antithetical to the desires of workers who wanted to inhabit a peaceful society. The Chicago anarchists' lauding of dynamite in the "radical press" ceased in the post-Haymarket backlash, but anarchism's reputation was already damaged.[86] Anarchism became labeled as a political movement that is violent above all else.[87] Anarchism's association with violence seemed so contrary to popular opinion that, a generation later, anarchist agitator Emma Goldman's denouncements of violence failed to find widespread public acceptance.[88] Thus, by turning possible supporters into detractors, preaching dynamite helped defeat anarchism rather than defeat capitalism.

In the end, a remarkable product of the Haymarket incident was the stabilization of preaching dynamite as authorities exploited the fear of dynamite to further suppress workers, and anarchism's enemies employed one of Spies's rhetorical tactics, *antistrephon*, to turn dynamite against its most vocal advocates. New York labor journalist John Swinton noted that "enemies of the labor movement" could turn the Haymarket dynamite bomb into a potent symbol to deride any type of worker activism as dangerous and unappealing.[89] The police, the mainstream press, and the government did just that, using the potency of dynamite symbolism to build and maintain public opinion against the accused bomb conspirators. Establishment newspapers reported that "every" anarchist protester carried nitroglycerine and that random citizens uncovered dynamite bombs under wooden sidewalks all over the city.[90] The Haymarket bomb gave the government the justification it needed to make examples of the anarchist press's workers, writers, and editors.[91] According to Chief of Police Frederick K. Ebersold, police investigator Captain Schaack "wanted to keep things stirring. He wanted bombs to be found here, there, all around, everywhere."[92] Police interviewed at Haymarket after the bombing showed no remorse for their attack on the rally and its participants, who had been "preaching dynamite for years."[93] The *Chicago Times* called for the extermination of the "slavic [*sic*] wolves" who perpetrated the bombing to avenge the dead policemen.[94] Spies reflected from prison that the "starspangled, shooting and clubbing votaries of Liberty, while searching for dynamite in private houses told the people that they just wanted to find 'enuff of doinemoit to blow that———Spies up with.'"[95] For such reasons, Schilling, in the same letter to Lucy Parsons, asserted that "the revenge circular of August Spies was met by the revenge of the public mind, terrorized with fear until it reeled like a drunken man, and in its frenzy swept away the safeguards of the law and turned its officers into pliant tools yielding to its will."[96] One dynamite bomb held the city hostage with the fear of more bombs, and only after Black Friday did the fervor in Chicago subside. After the trial, dynamite's ambiguous polysemy was reduced to an unequivocal symbol of anarchist terrorism, and its instability was brought under control. The Haymarket bomb transformed dynamite, as a metaphor for both general revolution and for the pinnacle of political thought in the hands of anarchists, into the literal representation of indiscriminate physical destruction and a disreputable metaphor for political naïveté. The exploitation of dynamite by the state and mainstream journalists established a durable, unappealing identity for anarchists as "bomb-toting, long-haired, wild-eyed fiend[s]."[97] The authorities had succeeded at turning dynamite against anarchism, and political stabilization thereby accompanied the rhetorical stabilization of preaching dynamite, at least for a time.

Notwithstanding the censure of violent language, Spies still turned to violence to agitate for anarchism by swapping the symbolic and material power of dynamite for that of the noose. Spies turned down a chance to ask for clemency, a chance that two of his codefendants, Fielden and Schwab, took. Instead, Spies embraced the opportunity to become a martyr for anarchism. "Call your hangman!" he insisted to Judge Gary at the end of his courtroom address.[98] He demanded to be executed to show that workers' preservation meant capitalist destruction and that workers' destruction meant capitalist rule. Spies hoped his death would hasten the destruction of the system that killed him. His ploy worked, to an extent. The executions provided the labor movement with four martyrs, which proved a longer-lasting symbol than dynamite was, and linked the names of Spies, Parsons, Fischer, and Engel with those of past and future American labor martyrs such as the Molly Maguires, Joe Hill, Nicola Sacco, and Bartolomeo Vanzetti.[99]

Spies understood that his unjust execution could become a type of embodied metonymy, representing the injustices perpetrated against workers worldwide and thereby revealing the truth of anarchist principles. "Truth crucified in Socrates, in Christ, in Giordano Bruno, in Huss, Galileo, still lives—they and others whose number is legion have preceded us on this path. We are ready to follow!" Spies declared in his final words to the court.[100] To tweak a concept from Kenneth Burke, Spies desired to become the "metonymic representative" of the whole anarchist movement.[101] Such comparisons to various historical martyrs were another commonplace that inflected Haymarket discourse. And just as these historical executions served as metonymies for oppressed philosophical, religious, and scientific ideals, Spies wanted his own death to function as an embodied metonymy for workers oppressed by capitalism. Martyrdom's significance thus derived from the capacity of one death to become amplified to such an enormous scale that it became representative of almost everyone's lives. The millions who witnessed his execution through the telegraphic, mass-media event magnified his death through memory and the repetition of observation rather than through throwing bombs.[102] His martyrdom used a moment of destruction to grant witnesses awareness, understanding, and insight about how to work for future preservation.

Beyond hoping that his execution might serve as an embodied metonymy for oppression, Spies also hoped that it would impel further anticapitalist activity. Spies stated that "the contemplated murder of eight men, whose only crime is that they have dared to speak the truth, may open the eyes of these suffering millions; may wake them up."[103] Only after the masses understood the injustice of his and his codefendants' martyrdom in relation to their own toil would they find the motivation to eliminate capitalism. Spies's violent demise provided "a symbolic means to transform events and ideas into a new rationale for human relations and collective action," to use the words of Stephen Browne's assessment of antiabolitionist violence.[104] Spies demanded death because it would hasten the destruction of the system that killed him. By embracing the court's judgment that he was the guiltiest party, he beckoned the hangman to ensure that his corpse, dangling beneath the scaffold, would embody the mass sacrifice necessary to spur revolution. He called the hangman to arms just as he called workers to arms, for both would serve the same purpose—advancing the demise of capitalism. The symbolic appeal of martyrdom thus had greater longevity than did the preaching of dynamite, even though dynamite survived its post-Haymarket-era calumniation to remain important to some anarchist and labor movements in Europe and the United States.[105]

4. Conclusion

Spies's execution for preaching dynamite conveyed to workers that their preservation or destruction rested in the control of technologies, whether in battle or in factories or in publishing houses. At Haymarket, technology influenced rhetorical invention, and rhetoric influenced technological invention. And the coupling of Spies and dynamite bomb, rhetor and weapon, language and object, empowered the strategic deployment of dynamite polysemy—in which the term could be simultaneously a literal call to throw bombs, a metaphor for revolution, an exhortation for self-defense, and a statement of historical inevitability—as well as paradoxical turnaround arguments in which dynamite proved that anarchist destruction would preserve the working classes from capitalist destruction. Moreover, this chapter's middle-way argument showed that the presence of dynamite in Chicago contributed to the trajectory of events, forming a resilient assemblage with anarchism, Spies, the courtroom, and the terrorized public mind, an assemblage that both displayed dynamite's force and impinged upon Spies's agential capacity as a rhetor.

The Haymarket events thus show how language spills over into the realm of physical destruction and how physical destruction spills over into the realm of language. Preaching dynamite was not just a manifestation of symbolic violence, but it was somehow constitutive of bloodshed in the inscrutable spaces between the various meanings of the word "dynamite," the rhetor Spies, and the weapon itself. When Spies preached dynamite, he operated within a militant discourse underpinned by anarchist political theory and the possession of dynamite bombs by revolutionary-minded workers. Dynamite not only infused Chicago's political discourse circa 1886 but also pervaded the Haymarket trial, both in language and in material presence. Spies confronted his audience with a serious choice: they either had to stand with Spies and the labor movement by admitting the fallibility and moral failings of a system that would hang innocent men, or they had to acquiesce to capitalist oppression. In this way, dynamite arbitrated peoples' responses to the political ramifications of Technē's Paradox.

In retrospect, current terrorism discourse reproduces the commonplaces of preaching dynamite, despite the waning importance of dynamite bombs. New weapons such as assault rifles, suicide vests, truck bombs, IEDs, and drones now provide polysemous meanings and opportunities for inventing turnaround arguments for terrorists and governments to exploit. Although IEDs and car bombs have replaced dynamite as terrorists' weapons of choice, and terrorists have replaced anarchists as the frightening enemy, contemporary weapons rhetoric shows that aggressors still mobilize unqualified high praise of weapons in order to capitalize on the paradox that technology will be what both annihilates and preserves populations. As capitalism, fundamentalism, tyranny, and democracy, locked in ideological disagreement, continue to generate wars and belligerence, mass media resound with talk of cowardly foreigners bent on annihilating civilization with ultimate weapons that, for the foreigners, mean preservation instead of destruction. The ongoing influential presence of weapons indicates that these rhetorical tactics will remain common.

3

Humane, All Too Humane | The Chemical-Weapons Advocacy of Major General Amos A. Fries

The inhumanity of [poisonous gas] is absolutely disproven by the results of its use in the World War.
—Amos A. Fries and Clarence J. West (1921)

Here is a killing instrument—gas—of power beyond the dream of a madman; here is a scheme of warfare which inevitably draws those who were hitherto regarded as non-combatants into the category of fair game.
—Will Irwin (1921)

When Amos A. Fries embarked for Paris and the Great War in late July 1917 as the director of roads for the American Expeditionary Forces (AEF), he thought he would be organizing a highway system for US troops and equipment.[1] He was, after all, a lieutenant colonel in the US Army Engineering Corps who had designed the roads at Yosemite Park and overseen other large projects, including the reconstruction of the Los Angeles and San Pedro Harbors.[2] But while Fries was crossing the Atlantic with the AEF, the Allies were scrambling in reaction to the German Army's newest introduction to World War I—mustard gas. Little could he have known that the US Army's unpreparedness for chemical warfare would bring him to proclaim the humane, preservative power of chemical weapons. Fries blew wind in the face of popular opinion about chemical weapons with a contrarian attitude that brought him face to face with Technē's Paradox.

In this chapter, I argue that Fries negotiated Technē's Paradox by categorically refusing to concede that chemical weapons could annihilate humanity. On the contrary, Fries declared chemical weapons to be categorically preservative. By remaining faithful to the pole of the Paradox that declared chemical weapons to be preservative, humane, and legitimate, Fries rejected the legitimacy of the other paradoxical pole that was upheld by his adversaries, who declared chemical weapons inhumane, illegitimate, and, above all, destructive. In this way, Fries did not traverse back and forth between the opposed poles of destruction and preservation inherent in Technē's Paradox like Thomas Malthus and August Spies did. Fries jettisoned technological ambivalence and uncertainty in order to take a firm stance in favor of chemical weapons as a preservative force.

As I turn from the previous chapter's examination of the Haymarket dynamite bombing to the rise of chemical warfare during and after World War I, I shift this book's focus from a somewhat isolated, less-destructive, and less-violent moment to a modern, globalized, and industrial war that reached an unprecedented magnitude of violence, horror, and weapons mobilization. The generations who witnessed World War I saw its generals fling bodies at the enemy in misguided assaults, the dismal slog of trench warfare, and the mass production of modern weapons such as tanks, machine guns, submarines, high-explosive (HE) shells, and, the focus of this chapter, chemical weapons. World War I manifested a "real orgy of destruction" as over ten million people perished, and "eschatological imagery took hold" of the world's population.[3] The magnitude of bloodshed attained in the war was a momentous innovation in warfare and the technologies of killing. And the violence of the war transformed the context of Fries's chemical-weapons advocacy—and modern-day weapons rhetoric in general—in several ways. First, weapons rhetoric was bound up with the fate of many more people than it had been during previous wars, so as much as the Napoleonic Wars amplified the size of standing armies and the scope of human misery in the early nineteenth century, the rapid European population expansion prior to World War I entailed ever-larger standing armies and the capacity to increase them further.[4] Second, ongoing innovations in transportation, communication, and propaganda technologies meant a much-larger audience existed for weapons advocacy. Third, World War I belligerents innovated vast centralized, economic, logistical, and organizational management systems to facilitate combat.[5] These factors have only grown larger in their purview since then, so understanding how weapons rhetoric worked at that time bears upon our understanding of how weapons rhetoric works now, at a time when our weapons far outstrip the destructive capacities of those used in World War I. The impetus to understand how the presence of weapons and words about weapons interact in the world's most violent conflicts remains urgent.

The severity of the first mustard-gas attack snuck up on its victims, and its battlefield results did not seem at all preservative. On the night of July 12–13,

1917, as Fries was preparing for his voyage, British troops of the Sixth Battalion's Fifteenth and Fifty-Fifth Divisions paid little attention to the shells that plopped into and around their frontline trenches. An unfamiliar, faint odor suggested a gas attack, but the lack of a visible cloud, a pungent smell, and physical symptoms indicated that the German barrage at Ypres had, aside from harassing the Sixth Battalion, failed. The soldiers thought they were duds. Still, having accustomed themselves to chemical warfare, many of the troops affixed their gas masks and awaited the all clear, which came the following morning. The fifty thousand shells were not duds. The German Army had saturated the front between St. Jean and Potijze with mustard gas.[6] As the day progressed, the battalion's five thousand soldiers developed large suppurating blisters, and their eyesight failed.[7] Many lay dead or dying. Over a million more German mustard-gas shells followed in the next ten days. Their two-thousand-five-hundred-ton payload doused the front. The number of British gas casualties spiked, but the British Army had already become accustomed to innovating defensive tactics and equipment rapidly.[8]

Unlike the other Allied powers that had dealt with chemical warfare since Germany's first successful chlorine attack in early 1915, the US military was thoroughly unprepared to defend against or use chlorine, phosgene, mustard gas, and other chemical weapons.[9] US casualties would spike, just as the British Army's had, unless the US Army organized a response. Enter Amos A. Fries. Instead of building roads, Fries was chosen to lead the US Army's entry into chemical warfare. Gen. John J. Pershing and his staff would not have known that when they selected Fries to organize the AEF's foray into chemical warfare, less than a week after Fries's mid-August arrival in Paris, they were about to promote a soldier who would approach chemical-weapons advocacy with as much zeal as he approached engineering projects. Within a day and a half, he had drafted an organizational paper and chart to deal with chemical warfare, and on his eighth day in France, he became chief of the AEF's Gas Service. In his new position, Fries had to learn a new field, figure out how to train green US soldiers in defensive measures, organize gas troops, procure chemical weapons and other *materiel*, and figure out how to use them. Over the next year, he built the Gas Service out of nothing, directing the somewhat stilted American response to chemical weapons in the chaos of adjusting to unfamiliar ways of waging a chemical war. In his words, the Gas Service "began in the field and was developed and operated in the field under the handicap of lack of knowledge, lack of production, lack of precedents and lack of supplies."[10] After President Woodrow Wilson's General Order no. 62 amalgamated various service branches in late June 1918 into a distinct bureau in the War Department called the Chemical Warfare Service (CWS), Fries became a US Army brigadier general and chief of the AEF's CWS.[11] Upon World War I's conclusion and after a temporary demotion caused by demobilization, Fries began building the CWS into a primary unit of the US Army, and in early 1920, he was promoted to major general and chief of the CWS. He maintained this position until his military retirement, in 1929, whereupon he became a virulent anticommunist and "right wing" education pundit dedicated to keeping military training in schools and communism out.[12]

Fries's military mission thus turned into a rhetorical one. Fries became one of the most outspoken advocates—a "super-military agitator"—for chemical weapons and the designer of a detailed chemical-warfare doctrine.[13] In order to justify the centrality of chemical weapons to twentieth-century warfare, during the war and after, Fries needed to develop offensive and defensive strategies, train battlefield chemists, and promote chemical weapons in the face of almost unanimous condemnation. More so than any other weapon deployed in World War I, gas received the most opprobrium, and its use stirred great controversy and debate owing to its novelty, its limited battlefield uses, the types of injuries it caused, and the magnitude of fear it generated. In the face of military and civilian denunciation, "Chemical Warfare Service officers have got to go out and sell gas to the Army," according to the slogan that Fries and West called the CWS's "watchword."[14] Fries proved himself a tenacious gas salesman. Anticipating that the future demand for "gas officers" would be "simply appalling," he mounted a profuse propaganda campaign to tout the necessity of maintaining chemical warfare to military command, soldiers, politicians, chemists, police, and the American public.[15] He wrote articles for military publications, chemistry journals, and popular magazines, and then he sent reprints of them—especially "The Humanity of Poisonous Gas"—to potential CWS advocates.[16] He wrote letters and provided direct-mail postcards to chemists in an attempt to enlist them to influence politicians, he gave speeches at military events, he testified before Congress, he massaged the egos of scientists, he arranged demonstrations of how to use tear gas for riot control for the Philadelphia and New York police, and he even wrote a negative book review of an anti-gas treatise.[17] Funding for these activities was provided in part by other pro-gas advocates in his circle.[18] He also promoted a CWS publicity stunt that, to tout the medicinal value of tear gas, oversaw the gassing of US senators, legislators, their staff, friends, and family, and he once tear-gassed his own daughter and dozens of other young girls.[19] Fries, the "super-military agitator," thus bore central responsibility for assuring both the presence of chemical weapons in American arsenals and the spillover use of chemical agents in everyday industrial consumer products.[20] In 1921, Fries published his most comprehensive statement on the subject, *Chemical Warfare*, cowritten with Clarence J. West.[21]

West's involvement with chemical warfare was a wartime episode of his career. Before the war, he had received his doctorate in chemistry from the University of Michigan in 1912 after writing a thesis on salt. When the United States began preparing for chemical warfare, the army, the US Bureau of Mines, and the National Research Council (NRC) organized about fifteen thousand American chemists to serve in either civilian or military capacities at home and overseas. West was enlisted, and during the war he served as the head of the editorial department of the CWS's Research Division. From 1918 to 1919, he worked on researching and testing gas masks at American University.[22] The CWS maintained close ties with the NRC after the war, working together on a number of joint projects, such as investigating how to use chemicals to protect pilings from marine borers.[23] West became director of the NRC's Research Information Service in 1921.[24] By and large, the publication of *Chemical Warfare* marked West's last direct involvement with the subject of chemical weapons, although as the director of the Research Information Service at the NRC, which served as an intermediary between university and government chemical-warfare research, he helped to keep track of American chemists and other scientists in case war should break out again.[25]

With *Chemical Warfare*, Fries and West needed to turn conscripts and officers into well-drilled battlefield soldier-chemists. Therefore, the book combined advocacy with pedagogical purposes, and much of it reads like a chemistry textbook. *Chemical Warfare* covers the full range of lachrymators (eye irritants), sternutators (vomiting gases), nerve agents, smokes, flamethrowers, and vesicants (blister agents) developed during the war and after, and how to utilize, research, and produce them. It includes a detailed guide to the chemical and technical manufacturing of mustard gas, complete with diagrams of requisite apparatuses and chemical formulas. *Chemical Warfare* also includes a history of chemical warfare, its development in World War I, and the formation of the CWS, in addition to covering offensive and defensive tactics and strategies. In terms of the book's broader aims, Fries believed that technical manuals on chemical warfare could convey the subject's "vital importance" to modern warfare.[26] Chemistry had become vital to military training, and World War I correspondent Carl W. Ackerman called *Chemical Warfare* "the first authoritative book on the subject of poison gas published in the United States."[27] And Fries and West's authority on chemical warfare lent weight to their claims about the humane, preservative capacity of chemical weapons.

The remainder of this chapter examines the chemical-weapons iteration of Technē's Paradox in three parts. The following section looks at the history of the chemical-weapons controversy and provides a middle-way approach to the debate about the humaneness of chemical weapons by using Kenneth Burke's concept of recalcitrance to examine what being in the presence of chemical weapons entailed. In section 2, a textual analysis of *Chemical Warfare*, I show how some of the book's rhetorical purposes—weapons advocacy, training soldiers, analyzing the tactics and strategies of chemical warfare, and making futuristic projections—helped Fries and West humanize chemical warfare by making it seem like a routine and unremarkable element of warfare, if not a benefit to everyday life. They established the preservative capacity of chemical weapons and endeavored to stifle the opposition with two prevalent rhetorical tactics. First, they advanced statistical proofs of chemical warfare's humaneness, and second, they amplified its humane, preservative efficacy to encompass the complete spatiotemporality of humankind. In the conclusion, I assess the results of Fries's publicity campaign with respect to the post–World War I history of chemical weapons.

1. Recalcitrance and Anti-gas Dissent

Much like Haymarket-era dynamite preaching, World War I–era chemical-weapons discourse was characterized by rhetorical instability. Chemical weapons in general, and mustard gas in particular, became polarizing, polysemous objects that could be mobilized for a wide range of incongruent persuasive goals. Yet, unlike Haymarket-era dynamite advocates—from anarchists to capitalists, and from journalists to factory workers—who all agreed that dynamite was a legitimate weapon, the participants in World War I–era chemical-warfare discourse became starkly divided over the legitimacy of chemical weapons. Although the discourse as a whole was paradoxical, few individual rhetors ever spoke of it as such. Proponents displayed unwavering devotion to the preservative power of chemical weapons, and detractors displayed unwavering, vitriolic condemnation of their destructiveness. Both sides remained adamant, and thereby the two positions together reiterated the paradox that what kills us will preserve us and what preserves us will kill us. In advance of the textual analysis of *Chemical Warfare* in the following section, this section traces the history of the humaneness stasis point, advances Kenneth Burke's concept of recalcitrance as another middle-way path that brings together the coinventive materiality of rhetoric and technology, and describes the anti-gas position that Fries and West attempted to counter as an example of recalcitrance in action.

In the decades before World War I, chemical weapons were a disputed issue in international peace negotiations and armaments conventions, and the 1899 and 1907 Hague Peace Conventions made the issue prominent. The problem of chemical warfare, as construed in diplomatic debates, revolved around whether killing with "poison or poisoned weapons"—rather than by steel or explosives—was legal. Both Hague treaties outlawed them. Article 23 of both the 1899 and 1907 treaties decreed that the "contracting Powers agree to abstain from the use of projectiles the sole object of which is the diffusion of asphyxiating or deleterious gases."[28] Germany signed the 1899 declaration, but the United States did not, and all of World War I's major belligerents signed the 1907 version. By taking a definitive stance against chemical warfare as illegal and uncivilized, the two Hague Peace Conventions nonetheless established the illegitimacy of chemical warfare as an explicit, but refutable, point of international law. At that time, though, no army possessed any viable chemical weapons, so the judgment about chemical weapons was speculative. After the war verified such speculation, the 1922 Conference on the Limitation of Armament clarified the reason for banning chemical weapons: they had been "justly condemned by the general opinion of the civilized world."[29]

Perhaps more so than the outright banning of poisonous gases in international law, the way certain diplomats framed the new weapons as humane indicated how chemical-warfare discourse would develop. Famed naval strategist and American diplomat Capt. Alfred Thayer Mahan provided the era's definitive statement on the humaneness of chemical weapons. He persuaded the 1899 American delegation to abstain from signing the Hague treaty because, he argued, asphyxiating gases were as humane as any other weapon: "The reproach of cruelty and perfidy addressed against these supposed [poison-gas] shells was equally uttered previously against fire-arms and torpedoes, although both are now employed without scruple. It is illogical and demonstrably humane to be tender about asphyxiating men with gas, when all are prepared to admit that it is allowable to blow the bottom out of an ironclad at midnight, throwing four or five hundred men into the sea to be choked by the water, with scarcely the remotest chance to escape."[30] This uncompromising proclamation equated poisonous gas's destructive capacity with that of conventional weapons. Its function, like any other means to victory, was to wreak maximum damage. In 1899, Mahan thus proclaimed chemical warfare to be neither humane nor inhumane. Rather, Mahan's statement implied that the objections to chemical warfare made by the Hague treaties and their advocates were inconsistent, even paradoxical, unless they dared to oppose warfare writ large. Nevertheless, he did give prominence to the question of humaneness, and this question became the defining stasis point of the ensuing chemical-weapons controversy.[31]

The rampant use of chemical weapons in World War I moved the debate from diplomatic circles into full public view. The humaneness of mustard gas—and of chemical warfare in general—was one of the main points that opponents agreed to contest directly. Consider the debate's basic positions as laid out by pro-gas advocate and medical journalist J. B. S. Haldane, who drew from his experiences as a doctor and one of the British gas troops during the war. "I claim, then, that the use of mustard gas in war on the largest possible scale would render it less expensive of life and property, shorter, and more dependent on brains rather than numbers. We are often told the exact opposite, that it will make it more barbarous and indecisive, and lead to the wiping out of the population of whole cities," he wrote.[32] This controversy persisted and, seemingly, almost everyone who addressed the topic of chemical warfare weighed in to declare it either humane or inhumane.[33]

Kenneth Burke's concept of recalcitrance is a useful way to understand how rhetors framed the humaneness and inhumaneness of chemical weapons, and how the argument that they were inhumane proved to be the more resilient one. Recalcitrance, wrote Burke, is the "flow" of "discoveries" that someone apprehends from a particular "point of view" and that makes revisions "necessary by the nature of the world itself."[34] In short, recalcitrance "may force us to alter our original strategy of expression greatly" in order to be believable and persuasive.[35] The recalcitrant rhetorical presence of chemical weapons functioned, in Burke's words, by substantiating, inciting, and correcting paradoxical claims about their destructive and preservative characteristics.[36]

Burke proposed two forms of recalcitrance, poetic and mechanistic. But mechanistic recalcitrance best addresses the question of what was entailed by being in the presence of chemical weapons. The two types of recalcitrance compel different but intertwined types of revisions. In order to function, technology, such as an airplane or artillery shell, must conform to some form of recalcitrant materiality, such as gravity, wind, propulsion, structural stability, and a clear flight path. Yet rhetors need not mechanistically conform to technological recalcitrance. People advocating, dissenting against, and arguing about machines and systems may use their poetic license, which is not immune to the recalcitrance of the poetic variety. Their material constraint might be technological, or scientific, but it is more likely to take the form of constraints imposed by ideology, politics, culture, and aesthetics. Technological recalcitrance makes it difficult for rhetors to lie outright about technologies and their effects, yet it empowers plenty of factual wiggle room for describing technologies, as Gordon Mitchell's examination of missile-defense advocacy well demonstrates.[37]

The famous description of the Titanic exemplifies the intermeshing of mechanistic and poetic recalcitrance. The Titanic's construction, appearance, and cost and the need to sell berths incited and substantiated its description as being "unsinkable." Mechanistic and poetic recalcitrance were in alignment upon the ship's embarkation. After the ship sank, the "unsinkableness" of the Titanic revealed that the description needed revision. Mechanistic recalcitrance caused the necessity of rhetorical correction, yet the ship's unsinkableness remains free to be used for ironic effect, being unconstrained by poetic recalcitrance. The Titanic checks our hubris and compels us to refrain from labeling any other seagoing vessel as unsinkable. The idea that any technology is "failsafe" is a fallacy, no matter what the design.[38]

Steve Fuller identified a type of "misdescription" that he called "strategic opacity," which helps to explain recalcitrance further. According to Fuller, strategic opacity is "a literal misdescription that is nevertheless necessary for the audience to act in a normatively desirable manner. . . . If strategic opacity succeeds, then in the long term the world comes to resemble more closely the strategic misdescription."[39] Put in terms of recalcitrance, strategic misdescription can persuade as long as an audience does not know or does not care that the description misleads. The recalcitrant power to necessitate revised and corrected descriptions always lurks nearby, however, and people can have a jarring realization if recalcitrant materiality exposes a misdescription as a falsehood. As the experience of World War I piled up evidence of the effects of chemical weapons, statements about them underwent recalcitrant revision. Battlefield data incited and substantiated the rhetoric of gas advocates and gas detractors. And both positions used strategic misdescriptions to exaggerate the effectiveness of chemical weapons,[40] yet over time the mechanistic recalcitrance of mustard gas, despite the more benign physiological and psychological effects of smoke-screens and teargases, tended to bend the idea of chemical warfare toward vilification and condemnation. Recalcitrance solidified the anti-gas point of view and undermined the pro-gas point of view, exposing the dogmatism of those, like Fries, who refused to give enough persuasive credit to the physiological and psychological horror of the new weapons.

In sum, both the anti-gas and pro-gas points of view found resistance to their misdescriptions of chemical weapons at the material level, but recalcitrance tended to substantiate the inhumane point of view more than the humane point of view. Although mechanistic recalcitrance is less pliable than poetic recalcitrance, both types of recalcitrant "inventions . . . follow until [a] point of view finds embodiment in our institutions and our ways of living."[41] According to Burke, statements that undergo a series of recalcitrant revisions until some sort of agreement is reached formulate "the public architecture of a social order."[42] After World War I, public opinion rejected the military use of chemical weapons such as chlorine, phosgene, and mustard gas, but barely registered the military origins of tear gas and validated the influx of toxic chemicals into household and industrial products.[43] Thus, the recalcitrance of chemical weapons substantiated a new way of living.

The rhetorical theory of recalcitrance advanced by Burke brings rhetoric into contact with STS scholarship about the construction of facts. When Thomas S. Kuhn pointed out in *The Structure of Scientific Revolutions*, a foundational work in STS scholarship, that the facts and theories established by science tend to be transient, alterable, and ever-changing, he initiated a long debate about the relationship of nature, objects, and observable facts to the creation of all kinds of knowledge. In the words of Langdon Winner, "It matters what a thing is, what name it has, and how people judge its properties."[44] And rhetoric, according to Thomas Farrell, can be considered "the fine and useful art of making things matter."[45] Often, as within chemical-weapons discourse, "what matters is constantly subject to negotiation, suspicion and revision."[46] Technologies matter. Weapons matter, and it is therefore important to measure words and objects, the names of weapons and the weapons themselves, and to make ethical and pragmatic judgments about them before settling on their facticity. Importantly, although Kuhn bristled in reaction to the rhetorical analyses he later encountered, *The Structure of Scientific Revolutions* also proposed that a rhetorical approach to science would help observers to understand how scientists—and, by extension, everyone else—create what we understand as reality.[47] Knowledge about how to act and rhetoric that incites actions are intermeshed. Rhetoric, in all of its mechanistic and poetic recalcitrance, does have an epistemological function that creates knowledge of the world.[48] Both points of view, pro-gas and anti-gas, were inventions both rhetorical and technological. And the reality that was apprehended, in all of its ethical, physiological, and psychological dimensions—the way people "defined reality"—depended on the interplay of words, objects, and interpretations.[49]

Words, statements, arguments, symbols, images, artifacts, and language in general persuade only if they undergo revision to conform to the recalcitrant image of the world that a society upholds, an image that itself is always under constant revision. Recalcitrance brings words and weapons together, sometimes in opposition and sometimes in alignment with public thought. By examining why chemical weapons like mustard gas have, in the dominant view, come to be known as inhumane, unethical terror weapons, and why other chemical weapons like tear gas and smoke screens have become banal, although contested, staples of military and police arsenals, we can begin to understand how weapons are recalcitrant, how they incite, substantiate, and correct statements about themselves. The use of mustard gas in World War I substantiated the inhumaneness of chemical weapons, which speculation about the weapons in military and diplomatic circles had incited. Gas's detractors focused attention on the recalcitrance of physiological and psychological trauma and horror. Chemical warfare was inhumanely destructive of bodies and terrorized the mind. Being in the presence of chemical weapons entailed that statements about the humaneness of gas, such as those of Fries and West, needed revision and correction.

Mustard gas exemplified the physiological effects cited by those who advocated the inhumane point of view. Mustard gas "burns the body inside or out, wherever there is moisture. Eyes, lungs and soft parts of the body are readily attacked."[50] Once absorbed into cells through various pores, mustard gas converts into hydrochloric acid through hydrolysis. The mustard-gassed body thus becomes a miniature acid factory and storage container for the toxic substance to burn, for perhaps weeks, until hydrolysis renders the gas inert.[51] Mustard gas therefore caused particular injuries that conventional weapons did not—burning from the inside out as well as from the outside in, prolonged asphyxiation from drowning on one's own fluids, and the development of putrid sores over the entire body. Although he wrote about the combined effects of various war gases, British soldier-poet Wilfred Owen described the experience of being gassed as "an ecstasy of fumbling . . . floundering like a man in fire or lime . . .

guttering, choking, drowning . . . white eyes writhing . . . blood gargling from the froth-corrupted lungs . . . vile, incurable sores on innocent tongues."[52]

The German Army's *Spezialtruppe für den Gaskampf* made sure that their enemies would feel these injuries. Victor Meyer had invented the German method of mustard-gas preparation in 1886, when he mixed thiodiglycol with phosphorous tricholoride. And when the *Spezialtruppe*, headed by Fritz Haber, discovered its usefulness as a weapon, the burgeoning German chemical industry already had the means to produce it in large quantities. Two companies, Bayer and C. A. F. Kahlbaum, manufactured the gas and filled shells at a peak rate of twenty-four thousand per day and two hundred thousand per month.[53] Once the German Army added HE to second-generation mustard-gas shells, their detonations disseminated the liquid in a fine mist that increased the likelihood of inhalation and skin contact.[54] These shells eliminated the telltale "plop" made by the original mustard-gas shells and made their explosive sound indistinguishable from other HE shells.

At first, avoiding injury during a mustard-gas attack was almost impossible when German mustard gas soaked the salient, especially in cold weather, when the gas's potency persisted.[55] As the German Army proceeded to launch as much mustard gas as it could, shelling the British trenches with it every night, the already-muddy battlefield became a "sucking quagmire . . . foul with every abomination."[56] In March 1918, witnesses reported that mustard gas flowed through the streets of Armentières like water.[57] After the air cleared, mustard gas still saturated the muck, threatening soldiers with incapacitating skin blisters should they slip into a crater. Casualties and fatalities mounted as soldiers, not appreciating mustard gas's persistence, handled tainted equipment, ate tainted food, touched each other, brushed the ground while defecating, and fell in the mud.[58] Nurses and doctors who treated the wounded became casualties.[59] Even after the infantrymen understood the characteristics of mustard gas, diving into craters to evade HE and shrapnel shells necessitated diving into mustard-gas-soaked mud against which their uniforms did not protect. To soldiers, this was "dirty warfare."[60]

Chemical weapons further corrected the pro-gas position by demonstrating that chemical weapons produced particular types of horrifying psychological damage. Chemical weapons triggered apprehension, panic, battle fatigue, and "gas mask exhaustion" like no other weapons.[61] Gas attacks demoralized both green recruits and veterans.[62] Frightened troops "stampeded" from the front.[63] Psychological horror, or "gas shock," accompanied the "frightful pandemonium" when a lookout cried "Gas!"[64] Recounting an early mustard-gas attack in his war diary, Hervey Allen noted that a new member of the British gas school "couldn't apprehend the fact that we [infantry] were suffering from the fear of gas, rather than from the gas itself."[65] Soldiers suffering from "gas fright" thought they had been gassed when no attack had occurred.[66] Another neologism, "gas neurosis," and the medical and psychological controversy about diagnosing "shell shock" were born from the atypical physical destruction of World War I's artillery bombardments.[67] Thus, owing to its physically and psychologically destructive power, mustard gas became a potent symbol of early-twentieth-century wartime destruction. It was World War I's most notorious chemical weapon, and it became known as the "king of gases."[68]

For the most part, gas substantiated the position of chemical warfare's opponents and American public opinion. Americans expressed a staggering dislike of gas and expressed an "outspoken and almost violent" hostility to chemical warfare, in the words of Fries and West.[69] An opinion poll about disarmament conducted in late 1921 tallied 366,795 votes for gas warfare's abolition and 19 votes in favor of its retention.[70] Another poll conducted by the US Department of State tallied 385,170 votes against gas and 169 for it.[71] Fries and the CWS asserted that "disinformation" about poison gas, such as Wilfred Owen's poem, had poisoned the public's knowledge of chemical warfare with inaccurate depictions.[72] No matter the source of negative public opinion, the public's stance was clear, and official US World War II military historians claimed that "there can be no doubt that gas warfare emerged from World War I with the reputation of a horror weapon even when field experience did not substantiate this view."[73]

From the anti-gas point of view, the prospect of waging chemical warfare with airplanes looked even grimmer than the results of World War I and incited a further correction of chemical-weapons discourse. Journalist Will Irwin's *The Next War*, also published in 1921, foresaw "a projectile—the bomb carrying aeroplane—of unprecedented size and almost unlimited range; here is a killing instrument—gas—of power beyond the dream of a madman; here is a scheme of warfare which inevitably draws those who were hitherto regarded as non-combatants into the category of fair game."[74] For Irwin, the prospective threat of sprinkling mustard gas from airplanes promised total war and global terror. To chemical warfare's detractors, the idea of more and more powerful chemical weapons was "monstrous and inhumane," in the words of Arthur James Balfour, British ambassador to the 1922 Conference on the Limitation of Armament.[75]

The Great War demonstrated that the world looked different through a fogged-up gas mask. C. K. Ogden and I. A. Richards, writing in 1923, called attention to the problem with the "conditions of communication" that well described 1920s chemical-warfare discourse. "The handiness and ease of a phrase is always more important in deciding whether it will be extensively used than its accuracy," they wrote.[76] The physiological and psychological destructiveness caused by chemical weapons proved paramount, providing just enough rhetorical recalcitrance for anti-gas detractors to prove the inhumaneness of chemically injuring and killing. The accuracy of calling chemical weapons inhumane proved "handy" for gas's detractors, whether or not the statement was accurate. But Technē's Paradox holds that the opposite argument can be made, and made with its mechanical recalcitrance. Fries, confronted by widespread reluctance, opposition, and hostility to chemical warfare, was adept at advocating for the CWS. As opponents and proponents swung the argument about chemical weapons back and forth, Fries attempted to halt the debate, to arrest and stabilize the reality of chemical weapons by defining them as irrefutably and humanely preservative.

2. Humanizing Chemical Weapons: Statistical Diminishment and Totalizing Amplification

After the Treaty of Versailles, chemical warfare was invoked for a broad range of political purposes, from arguing for pacifism to preparing for more war, while mustard gas symbolized the awfulness of war as much as it symbolized the imaginary promise of future weapons to abolish war.[77] Military command did not share the infantry's dread of the new weapons, but it did approach chemical warfare with trepidation. Fries and West wrote that during the war, "Much the hardest, most trying and most skillful work required of the Chemical Warfare Service officers was to persuade . . . Staffs and Commanders that gas was useful and get them to permit a demonstration on their front."[78] This reluctance remained after the war. So, in addition to facing civilians and soldiers sick of war and horrified by poison gas, Fries and the CWS also were unprotected in a time of great demobilization, army reorganization, funding cuts, and disarmament conventions.[79] At the time, Fries complained that the chief of staff and the secretary of war "seemed to be irrevocably committed to complete [*sic*] wiping out the service."[80] Hence, the stakes of Fries's public relations campaign were high. He was defending a novel but necessary military unit, a nascent but vast chemical industry, and attempting to ensure general military preparedness with respect to the looming communist threat.

Fries and West, seeking to dispel the image of chemical weapons as terrorizing, despicable, illegal, destructive, treacherous, and inhumane, attempted to stabilize the significance of weaponized gases by touting their effectiveness and viability as the primary weapons of choice. Hence, they confronted the Paradox as a distinctly polarized product of chemical-warfare discourse and adopted a suitable one-sided perspective. Fries and West were confident authors. "The Army and the general public have now so completely indorsed chemical warfare that it is believed the argument of inhumanity has no weight whatever," and the "inhumanity of [gas] is absolutely disproven by the results of its use in the World War," they declared.[81]

Categorically rejecting the claim that chemical weapons were inhumane, Fries and West made the case in *Chemical Warfare* for the power of gas to secure victory, maintain peace, reduce bloodshed, and preserve life with two primary rhetorical tactics.[82] First, they published casualty and fatality statistics to diminish chemical-warfare injuries in comparison to those of conventional weapons. In their view, the supposed trauma of chemical warfare was easily endured. Second, as they looked forward to the future of chemical warfare, the authors used amplification to depict how gas could penetrate the entire spatiotemporality of human existence. Starting by amplifying chemical weapons' lackluster battlefield records into a monumental achievement, they moved to their highest level of amplification—totalizing arguments that made mustard gas appear capable of permanently extending its humane, preservative power across the entire globe. With these two tactics, Fries and West thus attempted to stabilize the chemical-warfare controversy and settle it, unambiguously, in the CWS's favor.

In scattershot fashion throughout *Chemical Warfare*, Fries and West invoked statistics to demonstrate the humaneness of mustard gas and other chemical weapons. Whether they were describing a particular war gas, specific battles, and chemical-warfare tactics and strategies or advocating for these weapons, they used statistics to downplay chemical warfare's destructive power.[83] According to Fries and West, "The measure of humanity for any form of warfare is the percentage of deaths to the total number of injured by the particular method of warfare under consideration."[84] They hence measured the humanity of chemical weapons with statistics that contrasted the effects of conventional weapons with chemical weapons, made chemical-weapons defense seem like a banal routine, and educated soldiers about the necessity of defensive readiness.

Contrasting the casualty and fatality rates of chemical weapons with the rates of other World War I weapons provided Fries and West with a way to diminish the destructive power of chemical weapons by establishing their efficiency to injure but not to kill. More than just repeating a commonplace of chemical-weapons discourse, they implied that military commanders should embrace chemical warfare as a brand-new strategic method. An "official list of casualties in battle as compiled by the Surgeon General's office" cited by Fries and West showed that chemical agents caused 27.4 percent of 258,338 total U.S. casualties. Fries and West concluded from these numbers that "it is readily deduced that only 2 per cent of those wounded by gas resulted in death. That is, a man wounded on the battle field with gas had twelve times as many chances of recovery as the man who was wounded with bullets and high explosives."[85] By defining chemical weapons as injurious but not necessarily fatal, they situated war gases as simultaneously one of the least destructive weapons deployed in World War I and one of the most effective, since "no other element of war, unless you call powder a basic element, accounted for so many casualties among the American troops."[86] Unlike conventional weapons, they injured many but killed few. Chemical weapons thus provided the advantage of wreaking a much-lower amount of physical devastation than conventional weapons did.[87] Of the fact that World War I belligerents almost always used chemical weapons in conjunction with conventional weapons, Fries more often than not invoked his own observation that "silence is eloquent."[88] Fries and West's statistics indicated that military strategists should want to disperse enemy troops with harassment, light casualties, and little to no bloodshed.

This statistical argument seemed to overturn a basic principle of military strategy that armies must seek maximum destruction in war. Carl von Clausewitz asserted that the argument that one country could disarm another "without too much bloodshed" was a "fallacy that must be exposed," that "moderation" in war led to "logical absurdity," and that "the impulse to destroy the enemy . . . is central to the very idea of war."[89] Fries and West broke the connection between military success and destruction by implying that battlefield effectiveness derived not from killing and maiming, but from causing the greatest number of minor injuries. Considering their goal to "sell gas" to a military command that had been schooled in the art of causing massive destruction, the claim that chemical weapons were more useful because they killed fewer people and inflicted less-severe injuries than conventional weapons would have perhaps seemed a total reversal of war theory. They were proposing to win wars by preserving as many enemy soldiers as possible. In this way, Fries and West's statistical comparison of like phenomena purveyed chemical warfare as the humane way to wage all future battles in defiance of the quest for maximum annihilation.

Minimizing the destructive power of mustard gas served Fries and West's rhetorical goal of teaching wary soldiers that the weapon's killing power was easily surmounted with basic defensive measures. They wrote, "Due to the very slight concentrations ordinarily encountered in the field, resulting from a very slow rate of evaporation, the death rate is very low, probably under 1 percent among the Americans gassed with mustard during the war."[90] With this one-percent fatality rate, the authors conveyed that, of all the ways to die in battle, soldiers would probably not die from mustard gas, even after exposure. The one-percent fatality rate showed that mustard gas was less of a frightful terror weapon and more of a nuisance to endure within the more dangerous hail of steel. Fries and West did not deny that when troops were unprepared for an attack and ignorant of defensive measures, mustard gas's lethality increased. This lethality, though, was limited. In the greater context of their pedagogy, this statistic inculcated faith in defensive gas equipment and procedures. By knowing the simple defensive tactics that Fries and West taught—keeping calm, donning a gas mask, disposing of exposed clothing, rinsing chemicals from exposed skin as quickly as possible, and avoiding gassed areas—soldiers could preserve their lives in a chemical attack. The dead were poorly trained and, therefore, statistically insignificant.

Educating soldiers and citizens about chemical warfare in order to disabuse them of their ignorance and cowardice was, according to Fries and West, one of their main persuasive goals, which merged with another goal—ensuring the survival of American soldiers.[91] Moreover, ignorance, not genuine controversy, was responsible for the debate about chemical warfare, and statistical comparison would educate the ignorant. They wrote, "*We believe that all opposition to chemical warfare to-day can be divided into two classes—those who do not understand it and those who are afraid of it—ignorance and cowardice.*"[92] But facts would dispel both, they claimed, and once enough information made its way to the public, public opinion would turn in favor of the CWS. With respect to their goal of protecting soldiers, Fries and West described their rhetorical task in terms that would also help disabuse recruits of fear and ignorance. A lack of knowledge and training in chemical warfare would be fatal to the US Army. They wrote, "From the standpoint of the man at the front the Training Division [of the CWS] is one of the most important [army divisions]. To him gas warfare is an ever present titanic struggle between poisonous vapors that kill on one side, and the gas mask and a knowledge of how and when to wear it on the other."[93] But this fear needed tempering. "While the importance of impressing upon the soldier the danger of gas was early appreciated it was deemed necessary not to make him unduly afraid of the gas," they wrote.[94] With so much anti-gas opinion based on hyperbolic fear and ignorance, they asserted that the public needed a way to "check the accuracy" of the chemical-warfare information they encountered.[95] Accuracy, they claimed, would educate and calm soldiers and citizens alike.

Statistics about gas casualties would teach soldiers to train well and would reveal their fears to be baseless. Chemical-warfare fatality rates were low, and its propensity to cause severe injuries was likewise minimal. The severity of most mustard-gas casualties, for instance, appeared slight in their estimation, because the typical exposure to it was very low. They wrote that "one part in 14,000,000 is capable of causing conjunctivitis of the eye and that one part in 3,000,000 and possibly one part in 5,000,000 will cause a skin burn in a sensitive person on prolonged exposure."[96] Hence, many mustard-gas casualties happened after brief contact with a miniscule amount of the substance—just enough to cause reactions. Fries and West further speculated that "probably the majority of burns from mustard gas arose from concentrations of gas consisting of less than one part of gas to five hundred thousand of air."[97] Low-dose casualties, such as those caused by exposure to gassed woods, dugouts, shelters, clothing, and skin-to-skin contact caused blisters, minor lung damage, and temporary vision loss.[98] These injuries caused by low-level exposure did not appear very hazardous, much less menacing, and the sheer statistical preponderance of nonsevere casualties refuted the claim that it was an indiscriminate terror weapon that killed everything within range. Mustard-gas exposure statistics were, by Fries and West's argument, definitive proof that the weapon was humane. Fries's chemical-warfare strategy therefore held that the weapon should secure victory by temporarily incapacitating the enemy with low doses of chemical agents, and not by indiscriminately killing them.

As Fries and West sought to build trust in their system of reducing American casualties, the statistical mean proved the long-term effectiveness of their defensive gas tactics. Focusing on the statistical mean discounted the significance of mass casualties as statistical anomalies and once again downplayed the destructiveness of chemical weapons. They conceded that American casualties from German gas attacks "fluctuated through rather wide limits. There were times in the early days during training when this reached 65 per cent of the total casualties."[99] However, the final tally helped to prove the preservative power of chemical warfare by jettisoning the most chaotic and unstable moments of World War I, especially the battles for which troops were unprepared for gas attacks (e.g., the German Army's first uses of chlorine and mustard gas). Such battles, they argued, provided little insight into the stable character of chemical warfare. "On the whole the casualties from gas reached 27.3 [percent] of all fatalities," they concluded.[100] Thus, instead of the statistical anomalies caused by the introduction of hitherto unexpected types of gas attacks, the important number was the whole number, the constant, unwavering statistical mean that proved that mustard gas would injure but not kill if the readers of *Chemical Warfare* heeded its exhortations to exercise calm diligence during gas attacks.

The World War I battlefield data accumulated by Fries and West demonstrated that chemical weapons produced a very low percentage of deaths and only minor injuries, and therefore the effectiveness derived, counterintuitively, from their capacity to preserve the lives of combatants. Whether or not one believes that this line of argument was one of many World War I–era "statistical perversions,"[101] Fries ensured that his statistics were as reliable as possible.[102] And Fries and West used their statistical evidence to craft an ideal image of well-trained and defense-minded soldiers who escaped injury—an image of fighting that depicted the banality of defensive preparedness rather than the horror of battle.

Fries and West's statistical argumentation may have indicated to soldiers and the public that the danger posed by mustard gas should not cause undue fright or dereliction of training. But after using so much space to diminish mustard gas's destructive power, Fries and West risked making the weapon no longer seem powerful enough to warrant manufacturing, storing, and deploying it to the professionals whose livelihoods depended on the capacity to wreak wartime destruction and devastate enemy populations. If HE bombs and shells, machine guns, and tanks were more destructive, then they would have seemed more desirable for further research and development, especially given the novelty of Fries's chemical-warfare doctrine. Thus, Fries and West bolstered their statistical proofs by touting chemical weapons as possessing an unequalled capacity to be deployed across the entire globe at all times.

Even though Fries and West asserted that chemical weapons were somewhat nondestructive and preserved combatants, they still needed to tout their effectiveness, which they did by amplifying the power and range of chemical weapons to attain victory and maintain power. The authors used different levels of amplified magnitude to describe how chemical agents, especially mustard gas, would prove useful everywhere and for all time. They concentrated their use of amplification on depictions of what future gas warfare would be like with respect to the nascent yet effective deployment of chemical weapons in World War I. Fries and West used three levels of amplification, each extending the threat of chemical weapons and mustard gas to encompass more and more geographical and psychological space, and climaxing in a totalizing generalization about the future of chemical warfare. The first level amplified mustard gas's efficiency on the battlefield, the second level augmented the first level by extending the weapon's range to the whole earth, and the third level swept every human peacetime activity into the purview of preservative toxic chemicals.

The first level of amplification built upon the reputation of mustard gas and amplified its strategic power to encompass all elements of battle. Fries and West wrote that the introduction of "the most valuable war gas known at the present time" was so momentous that mustard gas "changed completely the whole aspect of gas warfare and to a considerable extent the whole aspect of warfare of every kind."[103] Mustard gas changed "the whole" of warfare by requiring a complete spatiotemporal battlefield adjustment to the new weapon. Its presence necessitated constant wariness wherever mustard-gas shells might explode and had exploded, which was almost everywhere on the front. It could persist for months, whereas an artillery shell's effectiveness did not persist beyond the first blast. Fries and West added a second temporal dimension to their amplification when they guaranteed that "in the future large numbers of these [HE and mustard gas] shell will be used."[104] Not only would mustard gas's injuriousness persist, but so would its strategic usefulness. Mustard gas, according to this logic, would seep into the complete spatiotemporality of war, because all offensive actions would benefit from forcing enemies to don gas masks; to wear stiff, impermeable clothing; and to treat the entire landscape with chemical neutralizers for interminable periods.

Second, Fries and West further amplified the power of mustard gas by speculating about how chemical warfare would extend the battlefield to encompass the entire earth and, by extension, all people. "One can hardly conceive of a situation where gas or smoke will not be employed," they wrote.[105] For example, "The high explosive mustard gas shell, not only because of its persistency but because of its quick deadliness, can be fired singly and be depended upon to do its work wherever there be men or animals."[106] Amplifying mustard gas's potential to be deployed "wherever there be men or animals" extrapolated the dangers faced by soldiers in a battlefield gas attack to menace everybody. Fries and West declared that they did not recommend using gas against noncombatants, but they implied that civilians would need to learn chemical-warfare defense as much as soldiers would.[107] Future chemical warfare would be unhampered by most problems of mobilization and logistics, Fries and West inferred, for the battle zone could be anywhere, especially where people had less preparation and warning, which was away from battlefields.[108] By this logic, mustard gas thereby began to seem like a more and more viable, useful, and successful weapon, one that the US Army should stockpile. If chemical-warfare preparedness would dictate military behavior, then in the future imagined by Fries and West, the global population must either wear a gas mask or face sure injury, albeit probably not a terrible one.

Fries and West, by amplifying the power of chemical weapons to threaten everyone at all times, brought gas advocacy close to total-war theory that aimed to militarize everyone. Fries and West concluded by concurring with Maj. Gen. William L. Sibert's preface to *Chemical Warfare* that "chemical materials as such become the most universal of all weapons of war."[109] Fries's 1921 *Annual Report of the Chemical Warfare Service* echoed total-war theory. He wrote, "The wars of to-day and those of the future will involve every activity of a nation and every inhabitant of that nation," a sentiment he echoed in his essay "The Future of Poison Gas."[110] Human research experiments and industrial accidents that threatened citizens far from World War I's front lines gave presence to Fries and West's second level of spatiotemporal amplification. But their positive assessment of this amplification ignored the ethical ramifications of promoting total war. By seeming to validate the ethics of total war but not the goal of seeking maximum destruction, Fries and West's rhetoric displayed what Richard Weaver called "melioristic bias." "Melioristic" refers to the idea that people can change the world for the better if they try hard enough. The "melioristic bias," according to Weaver, is "a deflection toward language which glosses over reality without necessarily giving us a philosophic vocabulary. One could go so far as to say that such language is comparatively lacking in responsibility. It is the language that one expects from those who have become insulated or daintified. It carries a slight suggestion of denial of evil."[111] The melioristic bias means that people tend to be so focused on how their own efforts benefit humanity that they are incapable of considering how much their own efforts might be harming it. No one would accuse Fries of being "dainty," but his advocacy for chemical warfare and denial of the charges of inhumaneness could be labeled "evil" by gas's detractors, pacifists, and other ethical positions that refute the morality of total war. His second-level augmented amplification of the power of chemical weapons to alter battle across the entire planet indicated that he had insulated himself from plausible counterarguments about the value of peace.

Fries was not against peace, though, and he very much wanted to avoid the bloodshed of a communist revolution, so it is essential to note how Fries's staunch anticommunism inflected his amplification of weapons at this second level. The communist threat that seemed to exist across the globe could be repelled with weapons with the same global reach—chemical weapons. Fries took a hardline stance against communism that included a zero-tolerance policy toward any group that threatened the post–World War I status quo in the United States. His conservative politics denied that any potential good could come from any organization that had any relationship whatsoever with a communist, a communist idea, antiwar sentiment, or any form of anticapitalism. "The communist is enemy number one at all times and at all places," he wrote.[112] He feared an American communist revolution, which would be a much more destructive war fought with conventional weapons rather than a preservative war fought with chemical weapons. For Fries, the communist "bloody civil war" or revolution would entail "*murder, assassination, wholesale slaughter of human beings*[,] . . . rape, arson, and general destruction of life and property."[113] This "bloody massacre" would be "the bloodiest kind of civil war," "the most terrible of wars," "**THE MOST CRUEL, MOST BRUTAL** and **FIENDISH CIVIL WAR IN ALL HUMAN HISTORY**" . . . "wherein neither age, sex, color nor physical condition will get any mercy," and it would result in the death "of hundreds of thousands, perhaps millions, of citizens."[114] In his book *Communism Unmasked,* he wrote that "the Communist Civil War . . . is the last act in the **BLOODY DRAMA** to be staged in every country."[115] Fries made sure his readers got his point: "**NO HELL ON EARTH**

HAS EVER EXISTED OR EVER WILL EXIST THAT IS A FRACTIONAL PART OF THE HELL OF A COMMUNIST CLASS WAR."[116] Such a revolution would be fought with steel and explosives, not chemical weapons, because the communists' aim would be total destruction of contemporary society. Thus, for Fries, defense against this communist threat that menaced the whole globe should be chemical, for humane chemical weapons could maintain the status quo without destroying anything other than communism.[117]

The universality of chemical weapons meant that they would help deter communism in a humane way, and they would help to preserve US government and society. Chemical weapons might promote "permanent peace," according to Fries and West.[118] In a 1921 article for *National Service* magazine, Fries wrote, "Perhaps the greatest guarantee that war will cease is the development of chemical warfare and the spreading of information that chemical warfare along with the Air Service will make war more and more universal, finally carrying it to the door of every citizen. When that day comes the world will see strenuous efforts made to settle disputes without resorting to war."[119] Fries and other chemical-warfare advocates thus wanted perpetual peace, but only a peace maintained by weaponized gases that threatened every person on the planet at all times.[120]

The third level of spatiotemporal amplification made not just war but every facet of human existence appear dependent upon one type of weapon. According to Fries and West, "gas is a universal weapon, applicable to every arm and every sort of action. Since we can choose gases that are either liquid or solid, that are irritating only or highly poisonous, that are visible or invisible, that persist for days or that pass with the wind, we have a weapon applicable to every act of war and for that matter, to every act of peace."[121] By proposing that chemical warfare would affect the global population's "every" activity in both war and peace, the magnitude of their amplification reached its uppermost totalizing limit. When they averred that peace and war would remain forever inseparable, Fries and West made chemical weapons seem unavoidable, necessary, and desirable. This speculative ideal applied to mustard gas as well. Fries tested the borders of the plausible when he expressed hope that a study showing that mustard gas retarded the development of tuberculosis in guinea pigs demonstrated that "there is a great field of usefulness for mustard gas from a prophylactic point of view."[122] For Fries and West, mustard gas and other chemical weapons had achieved what Thomas Hughes called "technological momentum," a state in which technologies become a "solution looking for a problem."[123] The "universality" of chemical weapons in peacetime broadened their appeal well beyond military affairs. According to this highest level of amplification—the "universal" application of chemical weapons in future warfare—they were absolute, predetermined, everywhere, unending, and inescapable. Fries and West thus established the categorical certainty that chemical weapons would remain commonplace, but this certainty need not cause undue fear and panic because the universal application of chemical weapons was preservative, not destructive.

This third universal level of augmented amplification tied into an important element of their public relations campaign that sought to justify the use of chemical agents at home, on farms, and in industry as even more beneficial during peacetime than during wartime.[124] Fries and West wrote that World War I's chemical weapons "are aiding to-day and will continue to aid in the future the peaceful life of every nation."[125] Amplifying the power of chemical weapons entailed a complementary amplification of chemicals in general. The "unlimited value" of this "unlimited field" meant that every nation would embrace the dual uses of dangerous chemicals in every action.[126] According to the rationale that chemicals "aid" humanity in all of its endeavors, the global population should welcome chemicals to assist with disinfection, water purification, pest extermination, dye-making, quelling riots, incapacitating criminals, hunting, doping airplane wings, developing photographs, perfuming, producing artificial fabrics, distilling gasoline, fertilizing, and curing the ill, for instance. The authors reconceived the potential humane extermination of wartime belligerents as using peacetime humane extermination for the preservation of humanity.[127] Fries and West thus granted chemicals a vast preservative power by framing them as essential in order to nurture and protect humanity from pests, disease, famine, and communists.

However, mass-producing chemical agents for all of these purposes meant that every nation would also be preparing for a massive chemical war. This military preparation also could preserve life through deterrence, if every chemical power possessed equivalent arsenals. This is the paradox of dual-use toxic chemicals; proof of the weapons' usefulness derived from the combination of peacetime and wartime functions. Fries and West thus used the material functionality of war gases as proof of the speculative adaptability of dangerous toxins to all human activities, and the speculative adaptability of chemicals to all activities proved the viability of maintaining the means to produce chemical weapons in perpetuity. Thanks in part to the advocacy of Fries, West, and the chemical industry, toxic chemicals would become commonplace in homes, businesses, and institutions. Extermination and preservation became "universal" synonymous activities, according to the logic of *Chemical Warfare*, when applied to, for instance, the use of chemicals to destroy agricultural pests. The individual soldier's problem, the "titanic struggle" between death and protection, became the world's problem and, by implication, so did the possibility of humanity's humane self-destruction.

Thus, Fries and West downplayed the destructiveness of mustard gas as a weapon by examining its statistical performance in World War I, and then amplifying its preservative power to a permanent, global scale. Despite taking a categorical stance in favor of chemical warfare's preservative power, the instability of chemical-warfare discourse slipped into the argument at times. The "gas is very deadly," they wrote of mustard gas in what seems like a direct contradiction of their statistical proofs that argued that mustard gas was not so deadly at all.[128] But wartime and peacetime equaled all time, and thus Fries and West advocated a permanent state of being in which chemicals presented everyone with the contrary prospect of inviting a weapon into the home to sanitize, exterminate, and grow food. "Notwithstanding the opposition of certain people who, through ignorance or for other reasons, have fought it, chemical warfare has come to stay," Fries and West concluded.[129] In *Chemical Warfare*, Fries and West showed that the anxiety produced by mustard gas and its destructive power never quite caught up with the hyperbolic rhetoric used to describe it. Like other chemical weapons, the effectiveness of mustard gas decreased throughout the war with innovations in defensive training and equipment, just as their statistical arguments bore out.

3. Conclusion

In *Chemical Warfare*, Fries and West's pedagogical goals, advocacy goals, and strategic goals resulted in a rather complex task as they confronted their central rhetorical dilemma—the stark, paradoxical binary constituted by disagreement over the humaneness of chemical weapons. Advocating weapons is a tough persuasive task,[130] and in the end, Fries's publicity campaign had mixed success. Fries and West did not overcome the "psychology of gas training," which, in their words, "they never succeeded in fully solving" for the soldiers whom the CWS inculcated into the chemical-warfare discipline, much less for military command, politicians, and the public.[131] In a famous example of continued opposition to gas, President Franklin Delano Roosevelt, when he vetoed renaming the

CWS the Chemical Corps and making it its own army branch, wrote, "It has been the policy of this Government to do everything in its power to outlaw the use of chemicals in warfare. Such use is inhuman and contrary to what modern civilization should stand for."[132] Despite Fries's confidence, his rhetorical tactics of diminishing the physiological and psychological harms of war gases with statistics and amplifying the range and uses of them to encompass the entire globe did not overturn anti-gas opinion. Being in the presence of chemical weapons and their recalcitrant materiality entailed the persistence of arguments that asserted chemical weapons' inhumaneness.

But the CWS remained intact, although it was often underfunded and understaffed compared to its World War I heyday. The massive chemical-weapons factory and proving grounds at Edgewood Arsenal often sat idle. Fries, though, was not an "unimaginative man" who could only devise an "unimaginative doctrine," like those in the World War I era who attempted to reconcile, for instance, machine guns and airplanes with outmoded military concepts.[133] Rather, Fries and West's foundational pedagogy helped to train soldiers as the CWS continued both to research new weapons, tactics, and strategies, and to develop elaborate drills and instructions for troops as part of a "comprehensive [military] doctrine"[134] to be taught in the CWS's gas school. Fries's tenacious rhetoric kept chemical weapons in the US military, and kept the chemical industry producing them. And by attempting to legitimate chemical weapons as distinctly humane and globally banal in their capacity to preserve life, Fries and West extended and added to the basic rhetorical tactics used by Malthus and Spies to negotiate Technē's Paradox.

The history of mustard gas between the world wars was also fraught with mixed results, as the weapon fell out of and into military favor.[135] For the most part, narrow deterrence, or the deterrence of a specific type of military action, worked, and chemical powers desisted from gassing each other.[136] Instead, they saved their chemical weapons for unprepared enemies and waged asymmetrical warfare. In the 1930s, the Japanese Army deployed mustard gas during its "Rape of Nanjing," and the Italian Army bombarded ill-clothed Ethiopian soldiers with it from airplanes. At intermittent times between world wars, the United States manufactured mustard gas at its Edgewood, Pine Bluff, Huntsville, and Rocky Mountain arsenals. Other countries, powerful and weak, did likewise, stockpiling chemical agents prior to and during World War II.[137] World War II commenced with the world's armed forces able but unwilling to wage chemical warfare. In battle, World War II belligerents ended up not gassing each other.

But within Nazi extermination camps, another of Fritz Haber's inventions—Zyklon B—continued to demonstrate the annihilating, inhumane power of chemical weapons. Meanwhile, a large-scale program of human mustard-gas experimentation in Australia caused many thousands of casualties.[138] Further medical research showed that chemical weapons are not as harmless in the long run as Fries and West argued.[139] The United States desisted from large-scale mustard-gas manufacturing in 1968, but the ease of producing it from chemicals readily available for commercial uses means that the threat persists. Thiodiglycol (TDG), a common precursor chemical used in mustard-gas production, requires just a one-step reaction—the addition of a chlorinating agent—to make mustard gas. Now, TDG is mass-produced for commercial uses and products, such as rubbers, lubricants, stabilizers, antioxidants, inks, dyes, photographic and copying processes, antistatic agents, epoxides, coatings, metal plating, textiles, solvents, cosmetics, and arthritis medication. TDG is more accessible now than during World War I and is, in terms of technological know-how, simple to produce in industrial quantities.[140] Owing to this availability of TDG and the ease of converting it into mustard gas, an international TDG black market has emerged.[141]

Fries anticipated this diffusion of mustard-gas know-how, if not who would possess it. He argued that mustard gas and other chemical weapons posed an internal threat to the US government if possessed by advocates of "the bloody terrorism of a Godless communism."[142] Mustard-gas production did not require a massive military-industrial system, and Fries understood that the CWS might not retain monopolistic control of the new weapons. He wrote that "the use of gas by unauthorized persons may prove a very serious problem. Practically all of these gases are so powerful that an ample quantity can be carried in a pocket to make it very dangerous in an ordinary room or even entire buildings."[143] By turning his attention away from foreign armies to armed individuals, Fries's rhetoric turned backward to the dynamite rhetoric that infused Haymarket discourse, in which a few pocketed bombs seemed to possess unlimited military power. In 1922, Fries even helped organize CWS troops to assault striking workers with gas, but they did not attack.[144] In fact, Fries and West had published a nearly complete how-to mustard-gas production manual complete with various chemical formulas, photographs of Edgewood Arsenal factory machinery, and detailed schematics of the German production method—information that could teach terrorists how to become battlefield chemists just as well as it could teach World War I US Army conscripts. In its most basic form, as a textbook in the vein of

Johann Most's *Revolutionäre Kriegswissenschaft*, *Chemical Warfare* could empower revolutionaries, terrorists, and less-industrialized states to manufacture mustard gas.[145] Fries thus recommended strict governmental oversight of chemical agents to keep the destructive power out of communists' pockets.[146]

Chemical warfare remains a subject of humanitarian concern and vigorous opposition, and despite the best efforts of Fries and West to stabilize into perpetuity chemical weapons as humane and preservative, the opposite pole of the chemical weapons Paradox has retained its recalcitrant prominence and stability.[147] Fries had the ears of generals, senators, and congressmen and the eyes of popular press magazine readers, and he helped the chemical industry burgeon. But he failed to convince the world that killing and injuring people with poison is beneficial.

As I write this chapter, the most notable usage of chemical weapons is in the three-way Syrian civil war among the entrenched Bashar al-Assad regime, the Islamic State (commonly known as ISIS and ISIL), and a motley array of fighters known as the Free Syrian Army. And a hundred years removed from World War I, judgment about chemical weapons still focuses on its inhumaneness.[148] Former US Secretary of State John Kerry might as well have been paraphrasing the 1922 Conference on the Limitation of Armament when, in his "Statement on Syria," he said, "It matters that nearly a hundred years ago, in direct response to the utter horror and inhumanity of World War I, that the civilized world agreed that chemical weapons should never be used again."[149] In contrast to Kerry's statement, Fries and West's conclusion to *Chemical Warfare* proclaimed, "How much better it is to say to the world that we are going to use chemical warfare to the greatest extent possible in any future struggle."[150] Not better, but worse, Kerry replied across the decades.

4

Toward a Peaceful Bomb | Leo Szilard's Paradoxical Life

Do not destroy what you cannot create.
—Leo Szilard (1940)

Readiness to accept the authority of science rests, to a considerable extent, upon its daily demonstration of power.
—Robert K. Merton (1938)

Leo Szilard's flash of insight in 1933 has become an atomic legend.[1] As the Hungarian scientist waited for a red light and crossed London's Southampton Row, his thoughts turned to a recent speech by nuclear physicist Lord Ernest Rutherford and to H. G. Wells's 1914 novel *The World Set Free*. Szilard pondered how to break apart atoms to produce vast amounts of power. Rutherford said it was impossible, and that to speak of controlling atomic energy was "talking moonshine."[2] Wells imagined a world rife with atomic power. In *The World Set Free*, the international proliferation of atomic power and bombs leads to global devastation and the subsequent creation of a world government. In Wells's doomsday vision, "the whole world was flaring then into a monstrous phase of destruction. Power after power about the armed globe sought to anticipate attack by aggression. They went to war in a delirium of panic, in order to use their bombs first."[3] Owing to the ease of obtaining the "simple apparatus," no institution could control the annihilation once it started. "The power of destruction which had once been the ultimate privilege of government was now the only power left in the world—and it was everywhere," Wells wrote.[4] So when Szilard crossed Southampton Row and conceived of how to create a nuclear chain reaction by dividing neutrons, he conceived of fission with a mental backdrop of massive destruction. Wells had taught him that the horrors of atomic war could overshadow the development of atomic energy, or make peace impossible. Still, Szilard became, by his account, "obsessed" with the idea of fission, and this obsession brought his life's activities into alignment with the destructive and preservative poles of Technē's Paradox.[5] One of his biographers characterized Szilard's "mode of being" as "science," but I prefer to characterize Szilard's "mode of being" as paradox.[6]

In this chapter, I argue that Szilard negotiated Technē's Paradox simply by being paradoxical. Whereas August Spies negotiated the Paradox by exploiting the rhetorical instabilities of language and weapons, and Fries and West attempted to stabilize them, Szilard lived a paradoxical life. Technē's Paradox influenced his activities, his rhetoric, and his technology as he placed himself at the center of the Bomb's crucial dilemma—whether a single technology could preserve or annihilate humanity. His biographical paradox is evidenced by his tenacious drive to develop, manage, and eradicate the Bomb.[7] This deep commitment to the most paradoxical of weapons dictated that his life and his words would reiterate and radiate Technē's Paradox with each twist and turn of the Bomb's early history. And as he advocated two contrary positions, he used two rhetorical *ethoi*—that of the disinterested scientist and that of the wily political advocate. One might call Szilard Janus-faced, but that would oversimplify the twists and turns that went into the invention and maintenance of his rhetorical being as he dealt with scientists, politicians, bureaucrats, soldiers, and the public. By describing him as a rhetorical being, I mean to conjure a conception of rhetoric that places triple emphases on the rhetorical invention of a person's persuasive means, social biography, and technological milieu. In the history of rhetoric, Quintilian's definition of the term is closest in meaning to what I intend to convey with the phrase "rhetorical being." For Quintilian, "rhetoric" meant a good person speaking well, and in ancient Rome, that entailed a complete training in both public oratory and a particular type of public morality.[8] So "being a good person" referred to conforming to a particular set of social norms, and "speaking well" referred to the capacity to be a persuasive public figure. Needless to say, Szilard did not conform to Roman societal standards, and he crafted his own globalized twentieth-century rhetorical being that incorporated his work on the Bomb as a vital component. His rhetorical being is observable as the total combination of his writings, his activities, and his technologies as they coalesced around the geopolitical moral ramifications of Hiroshima's and Nagasaki's obliteration. Inventing the means to use the Bomb for both preservation and destruction was his life's work. Szilard thereby lived Technē's Paradox.

Szilard was born in Budapest in 1898. At age twenty-four, he completed his doctorate in physics in Berlin. In 1933, he fled from the Nazis to Austria and then to England, and from England to the United States in 1938. As a scientist, he was more of an idea man than an experimenter, a "dazzling gadfly" flitting from lab to lab and suggesting ideas for others to bring to fruition.[9] In addition to brainstorming nuclear fission and critical mass, he made early, significant contributions to information theory, patented the cyclotron, and conceived of the electron microscope. After the war, he turned his attention from physics to biology and was appointed to the University of Chicago, where he stayed from 1946 to 1963. "Theoretically I am supposed to divide my time between finding out what life is and trying to preserve it by saving the world. At present the world seems to be beyond saving, and that leaves me more time free for biology," he quipped in a 1950 letter to atomic physicist Niels Bohr.[10] Later, he designed the radiation treatment for his own bladder cancer. He cured himself. Szilard spent his final year at the new Salk Institute for Biological Studies in La Jolla, California, an institution that he had urged polio researcher Jonas Salk to found. His biological research led to the development of the chemostat principle.

Szilard was also a political gadfly. His postwar peace advocacy included helping to pass the Atomic Energy Act of 1946 (also known as the McMahon bill), which granted control of atomic energy to a civilian group, the Atomic Energy Commission, rather than to the military.[11] He helped to organize the Federation of Atomic Scientists, the *Bulletin of the Atomic Scientists*, and the Pugwash conferences that brought together leading international nuclear physicists. He founded the political action committee Council for a Livable World, which still lobbies for arms control and raises campaign money for like-minded politicians. For this work, Szilard won the 1959 Atoms for Peace award alongside physicist Eugene Wigner. In 1960, he brainstormed the US-USSR hotline telephone connection between the White House and the Kremlin. Szilard persisted with his political interventions even as his professional stake in atomic power tempered his capacity to vilify atomic weapons, and his moral repugnance for the Bomb tempered his capacity to tout the weapon.

Szilard's Manhattan Project activities best exemplify his capacity to be at once scientifically and politically provocative and show how his central "obsession" goaded him to spend so much of his life developing and then attempting to control and eradicate the Bomb. With his early work on fission and his political foresight, he helped to spur the creation of the Manhattan Project.[12] Szilard's most notable writing was an August 2, 1939, letter sent to President Franklin D. Roosevelt, through Albert Einstein's mediation and with Einstein's signature. As a result, the president set up the Advisory Committee on Uranium (commonly known as the Briggs committee) to investigate nuclear energy's potential. Later, as an important member of the Project's Metallurgical Laboratory at the University of Chicago, he constructed and patented with Enrico Fermi the first atomic pile, or reactor.[13] He then turned his attention to electromagnetic reactor cooling systems. Szilard wanted to beat German scientists in the race toward the Bomb. Upon Germany's defeat, though, Szilard was horrified that the United States would drop the weapon on Japan and began organizing other Project scientists to oppose its use. Meanwhile, he also wanted to make sure that his work on atomic physics was rewarded, but by the time of the Trinity test, Szilard had little political leverage left to direct scientific and governmental policy, even as he clamored for it. Szilard's insistent attempts to influence policy-makers to refrain from bombing Japan and to negotiate patent rights with the government were regarded by Manhattan Project managers as meddlesome if not treasonous. The chief of the Chicago Metallurgical Laboratory, physicist Arthur Holley Compton, suggested to Gen. Leslie R. Groves, the military commander of the Manhattan Project, that the army should spy on Szilard's contrarian activities.[14] Groves thought Szilard a "villain" and attempted to have him jailed for the duration of World War II.[15] Groves conceded that the United States would not have produced the Bomb without Szilard's steadfast intervention, but the general declared, "as far as I was concerned he might just as well have walked the plank!"[16] At Groves's behest, the army's Manhattan Project security detail (and later the FBI and the House Committee on Un-American Activities) surveilled Szilard and his colleagues, whose policy and political ideas favored international civilian control of atomic know-how, all with little overt turbulence for Szilard aside from having his loyalty to the United States questioned.[17]

Through all these endeavors, Szilard found himself in paradoxical positions. Szilard depicted scenarios of atomic annihilation in order to make his research sound appealing to military and political leaders, yet he argued for a type of atomic preservation—the hope that atomic power might in some way deter aggression and sustain humanity with an unlimited energy source. Like other atomic scientists, Szilard's activities caught him hedging between preserving life and killing it.[18] Szilard believed that an American Bomb would create "a more livable world."[19] Yet he helped manufacture a more killable world.[20] He was thus paradoxical, as he simultaneously sought to create the most powerful weapon imaginable and to save the world from it. "Do not destroy what you cannot create," Szilard decreed in his personal "Ten Commandments."[21] But of the Bomb Szilard might well have commanded to himself: "Do not create what you cannot destroy."

To show Szilard's rhetorical being as he negotiated Technē's Paradox, the remainder of this chapter unfolds in three parts. In the first section, I examine his rhetorical being as it emerged from his writings and activities. I suggest that Szilard's rhetorical being, defined by the commingling of scientific disinterestedness and political interestedness, the commingling of secrecy and communalism, his embodiment of a unified scientific collective, and his atomic realpolitik attitude, demonstrates how he lived Technē's Paradox. In section 2's middle-way analysis, I examine Szilard's advocacy of minimal deterrence to suggest that the copresence of the Bomb and his words about the Bomb situated his rhetorical being on both of the opposed poles of Technē's Paradox. Different arrangements of scientific and speculative atomic facts entailed by the presence of the Bomb in arsenals and in fiction authorized Szilard's idiosyncratic deterrence advocacy that melded the preservative and destructive characteristics that the Bomb materialized. I conclude by examining Szilard's rhetorical being as a type of philosophic pragmatic idealism.

1. Technē's Paradox and Szilard's Rhetorical Being

In a high-stakes endeavor to devise wartime atomic policy, in which much chaos, complexity, and rhetorical instability competed with dogmatic ideological convictions and their accompanying rhetorical inflexibility, Szilard's rhetorical being reiterated a simple truth to soldiers, scientists, and policy-makers: they wielded the simultaneous capacity to annihilate humanity and to preserve it. Szilard was caught between two ideals, one scientific and the other political, which presented him with a difficult rhetorical task—convincing two disparate audiences about what, collectively, they should do about the Bomb. He wanted to persuade scientists that they should think and act more like politicians, and to persuade politicians to think and act more like scientists. When Szilard attempted to convince nuclear physicists to maintain inconspicuous secrecy about the likely destructive results of their research, provoked politicians to legitimate that research, and called for governments to abandon atomic weapons, his activities, always in the presence of the Bomb, combined opposed ideologies and politics, skirted the law, and birthed a peculiar rhetorical life. Szilard's paradoxical rhetorical being had three main characteristics. He crafted an ethos that borrowed from both science and politics, which he used to negotiate between the two fields of action; he embodied a peaceful, unified, and communal scientific voice in order to counter political enemies; and he crafted a realpolitik attitude toward the disclosure of atomic secrets.

"Ethos" is an Aristotelian term that refers to how the character of a rhetor makes that rhetor "worthy of credence."[22] Szilard had to make himself worthy of the credence of scientists at the same time that he had to make himself worthy of the credence of politicians. Scientific and political *ethoi* are dissimilar in their respective rejection and embrace of interestedness, or the willingness of people to act with partisan or self-interest. Scientists are supposed to be disinterested about everything but the discovery of scientific facts. Politicians tend to be self-interested partisans. Szilard, by combining scientific disinterestedness and political interestedness as part of his rhetorical being, was able to turn this ethotic dissimilarity into a tactical advantage.

As an ancient concept that still bears upon rhetoric, the concept of ethos warrants a further explanation of its contemporary relationship to science. The four normative characteristics of the scientific ethos that lend credence to scientists are disinterestedness (the elimination of nonscientific motivations from research and its uses), universalism (the assertion that observations and theories must match previous observations and theories), communality (the generation and sharing of scientific knowledge within a community), and organized skepticism (the suspension of judgment and the impartial examination of evidence until an informed judgment can be reached). These normative characteristics of the scientific ethos ideally secure political, social, and economic autonomy for the pursuit of science.[23] But along with scientific norms came ethotic counternorms. Scientific counternorms are interestedness (the appropriation of science for nonscientific and partisan uses), particularism (the assertion that observations and theories need not match previous observations and theories), solitariness (secrecy), and organized dogmatism (unwavering faith in a particular knowledge claim).[24] These scientific counternorms point toward how science becomes politicized. Both the norms and counternorms can coexist as potential elements of Aristotle's conception of ethos as the "fair-mindedness" of a rhetor at times when "inexact knowledge" leaves "room for doubt."[25] To build his ethos,

Szilard could have chosen "from among a range of strategic options those that [were] best suited to [the] situational contingencies," in the words of Lawrence Prelli.[26] Szilard manipulated the normative and counternormative characteristics of ethos by intermingling disinterestedness and interestedness, and communalism and secrecy, in order to adopt an ethos that, at times, was neither scientifically nor politically normative.

As a scientist, his exhortations in correspondence from 1935 and 1936 for the inner circle of physicists working on fission to display disinterest when sharing their research were not distinctive.[27] Regarding the theoretical-fission patents he developed from his London research, Szilard wrote to Rutherford in a May 1936 letter that "I cannot consider patents relating to nuclear physics as my property in any sense whatever. It would seem that if such patents are important, they ought to be administered in a disinterested way by disinterested persons."[28] On the surface, he was just being a typical physicist by researching the possibility of producing fission and repeating a scientific norm. Why he urged disinterest in fission, however, was conspicuous. Other atomic physicists, Szilard intimated, should aim to exude disinterest, for if their collective experiments succeeded and they failed to maintain strict control of atomic know-how, then they would introduce a world-destroying technology into a geopolitical context defined by the destruction wrought during World War I. Szilard wanted to remain aloof from the base motives attached to patents—proprietary control, financial considerations, and legal rights—lest the quest for registering patents appear mercenary when much larger financial, technological, industrial, and military considerations trumped individual financial gain. Throughout the late 1930s, Szilard pressed other atomic physicists to suppress their fission research in light of potential atomic weaponry.

Being scientifically disinterested, for Szilard, was more than just a question of individual accountability. He foresaw that by acting with disinterest toward patents and profits, atomic physicists might be able to exert some measure of control over the political appropriation of their science. Szilard did not want the technologies born from his insight to destroy the world, and he presumed his colleagues would concur. Szilard therefore, in short, proposed that by appearing disinterested, they could later act with interest. In March 1936, Szilard implied as much in a letter to Fermi. "I feel that I must not consider these patents as my private property and that if they are of any importance, they should be controlled with a view of public policy," he wrote.[29] Refusing compensation would dissociate their work from business, politics, and the military, even though

Szilard knew it would be bound to all three realms. Getting rich from their research would make nuclear physicists appear biased toward their monetary self-interest and undermine their policy proposals. They would be seen as military-industrial partisans rather than sage intellectuals. Szilard recognized that the most direct action that atomic physicists, as outsiders to politics, could take in order to gain political leverage over public policy was to eschew potential monetary gain from fission. Then, no one could accuse them of partisanship. The "disinterested control" over fission and fission patents that he recommended in a May 1936 letter to physicist John Cockroft was ironic in Szilard's intention to foster deep interestedness in his colleagues.[30]

Should eschewing financial gain not frame atomic physicists as viable policy consultants, the guise of "disinterested control" over patents promised another partisan usage to Szilard. He imagined that atomic physicists could use the proprietary rights granted by patents to force governments to desist from weaponizing their research. In the same letter to Cockroft, Szilard wrote, "If such multiple neutrons exist, we may envisage, if we wish to do so, the theoretical possibility of an industrial revolution in a not too distant future. In that case patents might be used by scientists in a disinterested attempt to exercise some measure of influence over a politically dangerous development."[31] In his estimation, Szilard and his colleagues could use their cultivated appearance of disinterestedness, via supposedly disinterested patent control, to mandate less-dangerous political uses of the Bomb. The disinterested scientific ethos backed any political and economic appeals that the atomic physicists might make. Fission was not just a matter of good science. Thus, Szilard's advocacy of disinterestedness was a ruse intended to carve out a political niche in which he and other atomic physicists could be very interested in the uses and abuses of fission by governments.

In addition to commingling disinterestedness and interestedness in his rhetorical being, Szilard commingled the scientific norm of communalism with the counternorm of secrecy. In a June 1935 letter to physicist F. A. Lindemann, Szilard recommended two principles for remaining objective toward fission: physicists should keep their research secret, only sharing it with trusted colleagues, and they should apply for patents only when they knew what their political or technological applications would be.[32] Szilard proposed a break with scientific norms by proposing to swap secrecy for the ethotic characteristic of communality. If physicists published fission theory too soon, then its industrial implications might not be apparent, which could stifle their careers. Paramount for Szilard's recommendation of secrecy to Lindemann, though, was Szilard's worry that a dangerous government, like Nazi Germany's, could appropriate it. The visions of H. G. Wells and forewarnings of Nazi aggression warranted a break in scientific protocol, at least within the small community of atomic physicists.

The ruse of appearing scientifically communal while being internally secretive was especially critical in the presence of atomic power. Szilard pursued atomic energy in a concerted and aggressive manner, and with full awareness of its weapons application. While the scientific ideal of utter impartiality remained idealistic, the military and financial ramifications entailed by releasing an inordinate amount of explosive energy did not seem to allow for much impartiality. But Szilard recognized that displaying and tweaking the prototypical scientific ethos could benefit physicists by later situating them as the best arbiters of how to use the products of their own invention, whether for annihilation amid the chaos of war or for monetary gain in a stable peacetime economy. Faking a disinterested ethos and swapping communalism for secrecy were, in part, rhetorical tactics meant to pay off later with either political leverage regarding atomic weapons or economic leverage regarding atomic power. As the Bomb moved from being a theoretical possibility to an imminent one, Szilard became disquieted. He could not maintain a guise of disinterestedness. The impartial attitude on the part of atomic scientists that he had hoped would position him and his colleagues to save the world from ruin began to appear more and more precarious, as was his ethical position as a Bomb physicist. With the annihilation or preservation of humanity in the balance, Szilard's 1930s version of the atomic-scientific ethos was no longer rhetorically viable even as a ruse. Szilard hence abandoned scientific secrecy to build a large community of scientists.

When science confronted the Bomb, Szilard argued that it should do so as a unified body, a body not only certain of atomic facts, but a body certain of its political convictions. Only the unified voice of the scientific community offered the most knowledgeable voice for policy-makers to heed, he believed. So Szilard sought to solve atomic weapons control, proliferation, and escalation problems by throwing the massed weight of a large group of intellectuals behind his policy work. He asserted that a unified scientific front could underwrite policy with the truth of science so that the US government would have to reckon with a formidable dissenting body of scientific brains. Szilard's pursuit of a collective scientific voice of truth was rooted in his experiences in Europe and within the

Manhattan Project. As early as 1930, Szilard had strategized that a collection of elite, youthful thinkers both "religious and scientific" that he called the Society of the Friends of the Bund should convene to steer the course of European politics toward democracy.[33] Science was a key to political progress, he thought. Paradoxically, however, Szilard attempted to encourage scientists to use their professional authority to dissent against the Bomb of their own making. Implicit in this paradox was a sense of responsibility. The ones most responsible for creating the means of annihilation should be the ones to devise how to preserve humanity from annihilation. Szilard's plan thus seemed to embody his paradoxical rhetorical being in the scientific collective.

The Manhattan Project's compartmentalization of scientists, which prevented them from sharing their research with each other, undermined scientific communalism by enforcing the strict secrecy for which Szilard had previously lobbied. With the wartime weaponization of fission under way, Szilard abandoned his quest for secrecy and attempted to revitalize communalism within the Project. "The lack of adequate, direct and continuous contact between those who take these far-reaching decisions and those who have first hand information of our needs is obviously a fundamental difficulty which might ultimately prove to be fatal," he wrote to the chairman of the Manhattan Project Military Policy Committee, Vannevar Bush, in July 1942.[34] Szilard worried that the compartmentalization of research would both slow down the Bomb's development and block scientists from reaching consensus about what to do with it. In a February 1944 memorandum sent to Bush, Szilard complained that scientists could not "form a well-founded opinion and even if they individually arrive at definite opinions they are not able to put collective recommendations on record."[35] In turn, "there can be no judgment on which the administration can base sound decisions."[36] As compartmentalized individuals, Szilard indicated, atomic scientists had little to offer when compartmentalization inhibited their collective political power. Compartmentalization had already proven problematic, he noted in the same letter to Bush: "If there had been a mechanism for putting this collective opinion on record it would have been difficult for the authorities who were responsible for taking far-reaching decisions to make the mistakes which were made because those in authority would have been faced with the choice of following the collective recommendation of the scientists or taking the full responsibility of going against a practically unanimous recommendation."[37] The US government, however, did not want science to function as a voice of power, or to have its authority and responsibility undermined by the authority and responsibility of science. It wanted to keep the science secreted away. To no avail, Szilard, in an unfinished memo, attempted to diminish the need for secrecy and promote communalism as good policy by clarifying to "'the authorities' that *a large group* of scientists freely communicating information is *a minor* danger to security."[38] General Groves did not budge regarding Szilard's provocations. To protect military and governmental power over the Bomb's development and use, and to assure the success of the Manhattan Project, Groves continued to mandate the compartmentalization of scientists until the Project's disbanding and thereafter, leaving a vision of the whole "collective opinion" for a select few managers, including Bush, Groves, and James B. Conant, to construct. Still, Szilard persisted.

Szilard's failure to influence the Manhattan Project administration to stop compartmentalizing research and information did not stop him from continuing to assert that the collective recommendations of scientists were vital to any and all wartime policy decisions. As the Trinity test approached, Szilard moved from recommending collective action to organizing it. Szilard appealed straight to President Roosevelt. In a memorandum from March 25, 1945, Szilard constructed a collective voice for his colleagues who opposed deploying an atomic bomb.[39] "Many of those scientists who are in a position to make allowances for future development of this field believe that we are at present moving along a road leading to the destruction of the strong position that the United States hitherto occupied in the world," he wrote.[40] Instead of dropping the Bomb, science, embodied in the voice of Szilard, recommended military restraint as the proper course. Szilard described the dire scenario he wished to avoid for Roosevelt. Sounding a bit like Gen. Philip Sheridan's assessment of dynamite bombs and journalist Will Irwin's speculation about chemical air warfare, Szilard wrote, "Under the conditions expected to prevail six years from now, most of our major cities might be completely destroyed in one single sudden attack and their populations might perish."[41] The gravity of the situation should have, according to Szilard, motivated those with decision-making authority to pay heed to the collective body of "men who have firsthand knowledge of the facts involved, that is, by the small group of scientists who are actively engaged in this work."[42] By this logic, Szilard equated collective scientific knowledge with prudent political knowledge as the best knowledge of the Bomb. The atomic physicists were an exceptional group and advocated military restraint at this singular moment. Nobody else knew what they knew. According to Szilard, the US government could assure its preservation in the face of potentially cataclysmic destruction only by heeding the collective recommendations of these physicists, at Szilard's behest, not to drop the Bomb on Japan.

The Trinity test on July 16, 1945, heightened Szilard's sense of urgency about making the collective voice of the atomic scientists heard. The day after the first atomic explosion, Szilard drafted "A Petition to the President of the United States." Recollecting his impetus to write it, Szilard remembered that "I thought that the time had come for the scientists to go on record against the use of the bomb against the cities of Japan on moral grounds."[43] With President Harry Truman then in office, he reused some of the same appeals to rational-scientific consensus and morality that he had used in the Roosevelt memo for the petition's cover letter that exhorted colleagues to sign. Szilard wrote, "I personally feel that it would be a matter of importance if a large number of scientists who have worked in this field went clearly and unmistakably on record as to their opposition on moral grounds to these bombs in the present phase of the war."[44] The moral grounds, he suggested, were founded in scientific knowledge and responsibility. "The fact that the people of the United States are unaware of the choice which faces us increases our responsibility in this matter since those of us who have worked on 'atomic power' represent a sample of the population and they alone are in a position to form an opinion and declare their stand," Szilard asserted.[45] Szilard therefore defined the role of nuclear physicists as one of unequaled moral importance that carried with it a huge burden of responsibility. The petition's cover letter thereby advanced an embodied synecdoche in which the scientists represented the whole of the US population and reminded the president that they, too, were American citizens or residents. If the average citizen had no political agency to alter Bomb policy, especially amid the ultrasecrecy of the Manhattan Project, then the few citizens and European exiles who knew were beholden to pressure the government for restraint and lucid deliberation about future consequences. Szilard thus took it upon himself to "form an opinion" for "we the undersigned" atomic scientists.[46] Of the hundreds of scientists employed by the Manhattan Engineering District, only sixty-eight scientists signed.[47] The petition never reached President Truman, whereas the Science Panel's report to Secretary of War Henry L. Stimson—signed by E. O. Lawrence, Compton, Fermi, and Oppenheimer—that recommended the military use of the Bomb on Japan quashed the chance that the Bomb might display its power in a nonmilitary technical demonstration.

When Truman authorized dropping Little Boy and Fat Man on Hiroshima and Nagasaki, respectively, power once again seemed to rebuff the voice of communal scientific truth as much as it was embodied in the rhetorical being of Szilard, who nevertheless remained resolute that the atomic scientists' collective opinion could influence future Bomb policy. That politicians sometimes listened to Manhattan Project scientists and kept them "on tap," if not "on top," gave Szilard some hope.[48]

As Szilard tampered with the ideal scientific ethos and attempted to develop an embodied, collective scientific voice, and throughout his atomic activities, Szilard's rhetorical being exuded a type of atomic realpolitik. His atomic realpolitik united the most cunning aspects of scientific communication with the most cunning aspects of political communication to formulate the means by which to advocate how to use the Bomb in both scientific and political policy realms. It empowered him to advocate morality in atomic policy and freed him from moral constraints when attempting to have atomic policy instituted. Robert K. Merton wrote that "readiness to accept the authority of science rests, to a considerable extent, upon its daily demonstration of power."[49] With atomic scientists hovering on the fringes of atomic policy-making, Szilard performed Merton's observation with "a mild and legitimate conspiracy," daily demonstrating the power of atomic science in a bid to gain authority over atomic policy.[50] Thereby, his language was partly based in the pragmatic engagement with politics and partly based in the ideal truths revealed by science.[51] Szilard thus recognized that both politicians and scientists needed to rely on each other to devise atomic policy at a time when many politicians viewed outspoken scientists as meddlers.[52]

Szilard meddled. He incorporated the tactics of realpolitik—subterfuge, provocation, and threats—into his rhetorical being, at first, in order to navigate and influence the funding of his research. In his correspondence with atomic physicists in the 1930s, in addition to urging them to maintain a disinterested scientific ethos while pursuing what he hoped were their mutual political interests, Szilard exhorted scientists to keep their research secret by adopting a form of "private publication."[53] But with the scientific ethos's usefulness for gaining political clout looking more and more doubtful, Szilard devised that releasing his research might be another way to leverage a place for both him and his science in government circles. His push for funding became more urgent after his joint fission experiments with Fermi in 1939 and the knowledge that Nazi Germany was pursuing an atomic weapon. He faced an overwhelming rhetorical task as he, an unknown foreign exile, attempted to insert himself into American politics as *the* expert on the Bomb. Thenceforward, Szilard dropped the outward appearance of a disinterested scientist and became an impassioned advocate.

Szilard's first foray into realpolitik with the Roosevelt administration saw him implement a type of protonuclear blackmail to generate governmental interest in his research. He offered a risky type of atomic bargain, insinuating that he would release information to physics journals, other governments, and industries about how to produce fission and, hence, the catastrophic weapon, if the government refused to back his work. The implication of his threat was apparent: If the United States wanted to assure its own preservation in the face of a looming world war, then it needed to fund Szilard's research. If the US government would not fund him, then other entities, including Germany, could wield Szilard's fission theory.

On April 5, 1940, after his and Einstein's joint letter to President Roosevelt failed to produce immediate funding, Szilard wrote to John T. Tate, editor of the *Physical Review*, regarding the atomic-pile paper "Divergent Chain Reactions in Systems Composed of Uranium and Carbon." "I am anxious that this manuscript should not be sent to print until I have definitely heard from the Administration that there is no objection to its publication," wrote Szilard.[54] In fact, the government appeared to have zero interest in whether Szilard published the manuscript. Later, he reflected that "I had assumed that once we had demonstrated that in the fission of uranium neutrons are emitted, there would be no difficulty in getting people interested; but I was wrong."[55] Szilard hoped that by convincing Tate of the national importance of his research, he could use publication—more directly, the threat of publication—to push the administration to pay more attention. His dilemma was thus generating interest in his research when secrecy was a priority.

The next month, still waiting for the government to step in and fund a classified fission project, Szilard had Einstein voice his appeal. As an outsider who lacked a more straightforward means to make his recommendations known to the upper echelon of the US government, he had few options other than using Einstein's authority to obtain some political influence. In March 1940, Einstein wrote to Alexander Sachs, one of Roosevelt's scientific advisors and Szilard's and Einstein's presidential intermediary. Einstein wrote: "Dr. Szilard has shown me the manuscript he is sending to the Physics Review in which he describes in detail a method for setting up a chain reaction in uranium. The papers will appear in print unless they are held up, and the question arises whether something ought to be done to withhold publication. The answer to this question will depend on the general policy which is being adopted by the Administration with respect to uranium."[56] Opaque when taken out of context of Szilard's correspondence over the preceding year, Einstein's reference to uranium policies alluded to Szilard's initial attempts to have the United States intercede in the world uranium market to scuttle the German Bomb. Using Einstein as an intermediary, he wanted to provoke the US government to act on the discovery of fission by initiating the foreign policies necessary to guarantee that the United States, instead of Germany, would secure uranium deposits. Germany had already seized Czechoslovakian uranium mines and could take over large deposits in the Belgian Congo if and when Germany invaded Belgium. However, Szilard's appeal, as conveyed by Einstein's letter, elucidated his grander ambitions. He did not only want to control the mere documentation of the first atomic pile and influence the uranium market. He was interested in more than "holding his papers up" and secrecy. Szilard wanted to move from being a scientific expert to a policy advisor. This ploy worked in two ways, first when the Briggs committee granted a modest $4,000 for his and Fermi's experiments. Then in June 1940, the Office of Scientific Research and Development, under the directorship of Bush, took over uranium research, which marked the first systematic stage of the US government's organization of the Manhattan Project.

Szilard's and Einstein's pressure had secured governmental interest in nuclear research, but in October 1940, Szilard thought progress was still too slow and funding inadequate, so he increased the intensity of his provocation. He wrote to George B. Pegram, chair of Columbia University's Physics Department, where Szilard was supposed to begin his research, to report that the US government had provided "no assurance that further support will be forthcoming."[57] Consequently, he suggested he might have to release secret nuclear patents to the Canadian and British governments: "If Dr. [Lyman J.] Briggs does not see any objection I might apply for a secret patent in Canada and in England, which I would assign without financial compensation to the respective governments, but I shall, of course, refrain from doing this unless I hear from Dr. Briggs that there would be no objection on his part or on the part of any other Government agency which is interested in this matter."[58] Short of handing reactor patents over to these foreign governments, Szilard proposed that if he still lacked "adequate government support," he would seek foundational support from Rockefeller or Carnegie, and that he was not above releasing the patents to industry.[59] With his open willingness to disperse secret documents Szilard increased the pressure on the Office of Scientific Research and Development to act before they lost control of classified weapons research during wartime. Whether a direct result of Szilard's gambit or not, the "Uranium Committee," then under the direction of Bush and the National Defense Research Committee, released $40,000 for Szilard and Fermi to build a uranium-carbon reactor. Thereby, his various attempts to bargain with the US government to pay attention to the threat posed by fission, to fund fission research, to pay attention to the world's uranium supply, and to maintain strict secrecy helped to materialize the first chain reaction that took place a year later at the University of Chicago. Szilard's wily and somewhat risky political interventions succeeded, in the sense that he had negotiated a small amount of political traction as he brought about the meshing of scientific, military, and political goals that coalesced as the Manhattan Engineering District in August 1942.

However, Szilard's later attempts to use his newfound realpolitik skills within the scheme of the Manhattan Project met with less success. Having already used his knowledge of fission as a political bargaining tool, Leo Samuel (Szilard's Manhattan Project code name) attempted to use patents derived from his work on reactor cooling systems to do the same with General Groves.[60] The "mixed feelings" Szilard experienced in 1942 regarding his work to develop the Bomb and the desire he expressed in a letter to Compton "to be sure that considerations of a purely financial nature played no part in determining my attitude" shifted toward the establishment of his patent rights.[61] The Manhattan Engineering District maintained strict control over all patents, and General Groves locked Szilard out for most of 1943 as he annoyed his supervisors with endless quibbling about the status of his patents.[62] Bush suspected Szilard of careerism, writing to Conant in January 1944, "I think Szilard is interested primarily in building a record on the basis of which to make a 'stink' after the war is over."[63] After the war, Groves said Szilard was "the kind of man that any employer would have fired as a troublemaker," and, "What a pain in the neck that Szilard was!"[64] Regardless of the enmity he garnered from his strong-arm tactics with Manhattan Project managers that, in part, contributed to his being surveilled by the government, Szilard had used the rhetorical tactics of realpolitik to strike a bargain that resulted in the creation of the Bomb.

Szilard continued to mingle the ideals of scientific communication with the realpolitik practices of political communication after he left the Manhattan Project and began lobbying for nonmilitary oversight of atomic science, disarmament, and minimal deterrence. During the Cold War, Szilard gave up neither his efforts to stem the arms race nor his efforts to unite a body of scientists to fight the unethical uses of atomic energy betwixt the threat of atomic war and the promise of nuclear deterrence. Thus, Szilard's work on fission in the laboratories, his tweaking of the normative scientific ethos, the embodiment of his policy recommendations in a unified scientific collective, and his realpolitik intrigues helped ensconce Technē's Paradox as a bureaucratic and political problem at the highest administrative levels of governance.

2. The Presence of the Bomb and Its Facticity

This section turns to the policies Szilard advocated, and how the copresence of the Bomb and his words about the Bomb situated his rhetorical being on both of the opposed poles of Technē's Paradox. Szilard did not exhibit any particular faith that the Bomb would bring the destruction or the preservation of humanity. Both results seemed plausible. As he worked for world peace, he stayed prepared for disaster, never settling down in any one location for long, residing in hotels, and living out of suitcases, one of which was his personal archive.[65] All the while, Szilard dealt with the Bomb's dual facticity—its simultaneously explosive, radioactive reality and its apocalyptic potential. The Bomb's presence compelled Szilard to ask a question: how might "the world . . . learn to live for a while with the bomb?"[66] The answer, for Szilard, depended on different arrangements of scientific and speculative atomic facts entailed by being in the presence of the Bomb. This chapter's middle-way analysis argues that the fact of the Bomb in arsenals and the fact of the Bomb in fiction authorized Szilard's idiosyncratic deterrence advocacy. This section begins by examining Szilard's conception of facticity and how the flexibility of facts gave mutually reinforcing credence to Szilard's advocacy of minimal deterrence in deliberative policy writing and in science fiction.[67] The copresence of the Bomb and the apocalyptic atomic holocaust narratives that saturated popular imagination after World War II—both of which Szilard helped to produce—further imbued Szilard's rhetorical being with a paradoxical quality as he attempted to steer the world's atomic fate toward deterrence-based peace.

For Szilard, the plasticity of atomic-fission facts provided a way to negotiate the conflict between realpolitik and atomic research. In a 1943 conversation with physicist Hans Bethe, Szilard recalled saying, "I am going to write down all that is going on these days in the [Manhattan] project. I am just going to write down the facts—not for anyone to read, just for God." Bethe asked, "Don't you think God knows the facts?" Szilard replied, "Maybe he does . . . but not *this* version of the facts."[68] Szilard recognized that in the political sphere, he could rearrange facts in myriad ways to present competing arguments for both disarmament and arms reduction without losing sight of the scientific principles that undergirded fission. Scientific facts and political facts did not match each other. Rather than being a constraint when he attempted to have the government pay attention to his and the Federation of Atomic Scientists' policy ideas, the flexibility of evidence empowered both his pragmatic approach to arms reduction and his idealistic goal of disarmament. His self-proclaimed "addiction to the truth" should therefore not be confused with an addiction to factualness.[69] Controlling how policy-makers interpreted atomic facts thus became a primary concern of Szilard's interventions in Cold War policy. Aligning the facts of atomic research, national and international political goals, and weapons deployment could serve the truth of fission and ameliorate its devastating power. Perelman and Olbrechts-Tyteca wrote that "when facts are mentioned, we must always ask ourselves what it is that their use can strengthen or weaken."[70] Szilard did exactly that as he balanced scientific facts with speculative facts to strengthen short-term Bomb-control policies and long-term disarmament, and to weaken escalation. If he ever ran into trouble owing to his juggling of different versions of atomic facts in policy advocacy and fiction, Szilard knew he could always fall back on the record of his activities, both political and scientific, as historical facts. Hence, when in the short story "My Trial as a War Criminal" Szilard imagined the defeat of the United States in World War II and being prosecuted by Russia for his work on the Bomb, he envisioned that his best line of defense was to tell the truth by stating his activities without embellishment—that is, by stating his version of the facts.[71] Facts, Szilard thus recognized, could be rearranged to depict different realities depending upon differing scientific, political, and fictional exigencies.

Before I examine Szilard's mobilization of facticity's flexibility as a type of rhetorical tactic that he used to advocate minimal deterrence, it is important to note that Szilard's conception of flexible facticity contradicted the scientific ideal of what constituted a "matter of fact."[72] Drawing attention to the human control over and flexibility of scientific research could be considered one of STS's key disciplinary traits since, after the influence of Thomas Kuhn, much STS scholarship has aimed to reveal scientific and technological facts, knowledge, truth, and objectivity as either socially constructed or contingent upon human and technological vagaries.[73] Facts do not come from nature, science, logic, and the Bomb alone, so atomic facts can derive as much from cultural creations like fiction as they do from science.[74] Szilard surmised as much, composing articles for the *Bulletin of the Atomic Scientists* and fictional short stories about deterrence theory that demonstrated how he can be considered a disciplinary forerunner of STS who "questioned the ideals of knowledge" in post–World War II society, "pursued other paths to knowledge," and created the "new identity" of "the critical scientist."[75] As a "new kind of public servant" who combined scientific and diplomatic skills, Szilard mobilized the flexibility of facticity when he intervened in the controversy about how the United States should manage nuclear deterrence.[76]

Deterrence theory transformed the Bomb's paradoxical promise of global extermination and permanent peace—at least between atomic superpowers—into an ideology that underpinned Cold War logic. Military deterrence seeks to dissuade enemy attacks by promising annihilation as retribution, which, in theory, avoids either war or the use of a particular weapon. Deterrence assumes that the more devastating the threat, the more powerful the deterrent; consequently, deterrence can lead to weapons escalation. A holdover from criminal justice and World War I–era armament controversies,[77] deterrence became pervasive when, after World War II, states clamored to obtain the world-destroying power that they were powerless to use.[78] Nuclear strategist, RAND Corporation analyst, and hydrogen-bomb enthusiast Herman Kahn defined deterrence as using the Bomb to create a "*reliable* balance of terror."[79] Less confident in the prospects for successful deterrence than Kahn, nuclear strategist Albert Wohlstetter called it "a *delicate* balance of terror."[80] In Wohlstetter's estimation, "If peace were founded firmly on mutual terror, and mutual terror on symmetrical nuclear capabilities, this would be, as Churchill has said, 'a melancholy paradox;' none the less a most comforting one."[81] Cold War deterrence thus relied on the Bomb in order to simultaneously threaten humanity with annihilation and to preserve humanity from annihilation. Political philosopher Gregory S. Kavka asserted that "deterrence is a parent of paradox."[82] I suggest the opposite: Technē's Paradox is the parent of deterrence.

Alone, the destructive capacity of the Bomb provoked many paradoxical reactions, but once the Bomb united with deterrence theory, the weapon, in its ever-more-powerful designs, ensured the Bomb's prominence as the weapon that best exemplifies how Technē's Paradox has become ensconced in the collective global mindset.[83] Deterrence, a complicated balance of hope and terror, both wrought and imagined, functioned rhetorically by bringing the few atomic explosions between the Trinity Shot and 1963's Limited (atmospheric) Test Ban Treaty into alignment with the innumerable rhetorical bombs dropped by politicians, creators of pop culture, journalists, and rhetors of all kinds. This rhetorical alignment gave presence to the weapon and credence to words about the weapon. With each iteration, deterrence theory ensconced the Paradox as the nuclear superpowers' official foreign policy and dictated that humanity would be, for a time, dominated by the dual assurance that if deterrence worked, then humanity would abide, but if it failed, then humanity would commit collective suicide. Szilard, too, brought atomic fact into alignment with atomic fiction.

Both past atomic explosions and speculation about future ones warranted deterrence theory. Cold War deterrence was therefore proven by actual events, including Trinity, Hiroshima, Nagasaki, and the ongoing US atmospheric testing program that extended through the 1950s, as well as by events that had not, and might not ever, occur.[84] Deterrence thus relied both on the Bomb's presence and on the absence of nuclear war. This "tension between presence and absence in nuclear realism" as it related to deterrence was strong.[85] It underwrote paradoxical narratives about the Bomb's simultaneous capacity to annihilate—as evidenced by Hiroshima and Nagasaki—and, through deterrence, its incapacity to annihilate. About this incapacity, Jacques Derrida offered the infamous assessment that the "non-event" of nuclear war is "fabulously textual."[86] Physicist Edward Teller also took the stance that "there is no such danger" of "the mass suicide of the human race."[87] And Kahn took umbrage at how the hyperbolic magnitude of Bomb rhetoric did not match the magnitude of actual destruction the weapon could wreak.[88] The absence of a massive retaliation and a global nuclear holocaust, however, did not nullify the rhetorical presence of nuclear weapons in decades of nuclear testing by the United States and other nuclear powers, the persistence of radioactive fallout, the stockpiling of larger nuclear arsenals, and the pressure of the Bomb in politics. Speculation about doomsday, warranted by Hiroshima and Nagasaki, and speculation about disarmament thus provided an impetus for Szilard and others to advocate deterrence as a practical means to manage the Bomb.

Long after atomic scientists and sci-fi fans knew what the presence of the Bomb would entail but three years before the Trinity Shot, Szilard was motivated to put speculative atomic facticity to use with an appeal to fear in order to advocate deterrence. Like that of a number of his colleagues, Szilard's advocacy for the Bomb was rooted in anxiety about the wartime intentions of Germany.

He argued that the United States should rush to construct the Bomb to deter a catastrophic German offensive. "Nobody," he wrote in a May 26, 1942, letter to Bush, who was at the time President Roosevelt's scientific advisor, "can tell now whether we shall be ready before German bombs wipe out American cities."[89] Germany would not possess the Bomb. Nonetheless, in line with deterrence theory, he envisioned that obtaining this most violent device might temper the extent of Germany's immediate war effort.

Although Szilard held that atomic deterrence could work in the short term with respect to Germany, he predicted that deterrence would fail to maintain peace in the long term. In subsequent 1942 correspondence to Bush and other Manhattan Project managers he warned, "What the existence of these bombs will mean we all know. It will bring disaster upon the world even if we anticipate them and win the war, but lose the peace that will follow."[90] Losing the peace would entail atomic annihilation since, as Szilard asserted, a deterrent based on the United States' initial monopolization of Bomb technology would do little to stop countries from pursuing their own nuclear weapons programs. Rather, as the unequivocal harbinger of global holocaust in the vein of Wells's *The World Set Free*, the presence of the Bomb in US arsenals would necessitate proliferation. With eventual weapons parity, Szilard envisioned, deterrence would not work to halt war when one belligerent could still wipe out another before they had a chance to respond in kind. The obliterating force of the Bomb might deter certain acts of violence, but wars would still be fought. Thus, his negative assessment of deterrence's viability was at the center of his apprehension about the Bomb's effects.

Still, by announcing the terrible facticity of the Bomb, the first atomic blasts might serve a rhetorical purpose by goading world peace, Szilard hoped. He proposed that by witnessing the Bomb in action, humanity would be obligated to deter similar levels of violence in the new "Atomic Age."[91] A demonstration of the Bomb's destructive capacity would rally public support for lasting peace by embedding a powerful sense of dread and foreboding into people's minds, Szilard asserted. In a January 1944 letter to Bush, Szilard wrote, "If peace is organized before it has penetrated the public's mind that the potentialities of atomic bombs are a reality, it will be impossible to have a peace based on reality."[92] Szilard waffled in 1944 and 1945 about whether the United States should destroy a populated or pre-depopulated city, but Bush ended up concurring that a destructive demonstration was necessary to scare people into political pragmatism.[93] Only by fearing the atom could citizens and politicians best choose how to control it. "It will hardly be possible to get political action along that line unless high efficiency atomic bombs have actually been used in this war and the fact of their destructive power has deeply penetrated the mind of the public," Szilard continued.[94] Irrespective of whether deterrence would succeed in the short or long term, the presence of the Bomb entailed that fear would dominate policy-making. But any peace brought by the Bomb would not last long, Szilard decided.

Rhetorically being in the presence of the Bomb entailed a fictional facticity for Szilard and for others. The annihilation of Hiroshima and Nagasaki had proven that the Manhattan Project had created a geopolitical situation with the potential to devolve into the violent chaos predicted in *The World Set Free*. The annihilation of cities the world over appeared to be becoming tenable if not inevitable. In a 1955 letter to the *New York Times*, Szilard invoked Wells's novel in the first sentence as an exhortation for restraint and peace.[95] "All the things which H. G. Wells predicted appeared suddenly real to me," Szilard recollected.[96] He got to work, using his scientific credentials to craft his rhetorical being into a political figure whose policy interventions would, in the words of Walter R. Fisher, perform a "public moral argument" that blended his technical expertise with the responsibility to steer humanity away from nuclear war. Fisher noted that "the expert," in this case Szilard, "assumes the role of public counselor whenever she or he crosses the boundary of technical knowledge into the territory of life as it ought to be lived."[97] Szilard could mobilize his technical expertise to arrange the facts about what had happened in the scientific and military past as well as to arrange the facts of what would happen in a future war—that is, an "imaginary war"—in order to avoid it.[98] In a sense, Szilard was organizing his own style of "apocalypse management" as a more stable antidote to governmental mismanagement of the Bomb.[99] Once the boundary between the technical and the public sphere is crossed, according to Fisher, "the public, which then includes the expert, has its own criteria for determining whose story is most coherent and reliable as a guide to belief and action."[100] Szilard's moral narrative of imaginary war was especially dire, and a few examples from his documents provide a sufficient sampling of the doom-speak he used to guide both belief in what the Bomb entailed and actions that might create "a more livable world."[101] Sounding like H. G. Wells, Szilard wrote in 1934 that "the discoveries of scientists have given weapons to mankind which may destroy our present civilization if we do not succeed in avoiding further wars."[102] In 1942, he predicted "a world in which a lone airplane could appear over a big city like

Chicago, drop his bomb, and thereby destroy the city in a single flash."[103] And in 1945, he predicted all economic powers could achieve Bomb know-how without much difficulty. "All the cities of the 'enemy' can be destroyed in one single sudden attack," while "all of our major cities might vanish within a few hours," with the result that "the cities of the United States as well as the cities of other nations will be in continuous danger of sudden annihilation," he wrote.[104]

For politicians and the public, Szilard conformed to the rational demands of his narrative as he balanced doomsday prophecy with atomic hope. He was met with mixed results. The drawback of advancing a public moral argument about atomic doom based on scientific expertise was that, as a scientist entering into politics, his "moral concern" ended up being subverted by the juxtaposition of his technical expertise with his apparent political inexperience.[105] Nonetheless, his depiction of imaginary war possessed facticity after Hiroshima and Nagasaki because, in the words of Perelman and Olbrechts-Tyteca, he confronted his readers with a scenario for which he could "postulate uncontroverted, universal agreement with respect to it."[106] And after Japan's "unconditional surrender" and the end of World War II, Szilard's authority as an atomic scientist empowered his rhetorical being as he mobilized the Bomb's factual-fictional copresence as a pliable tool in domestic and international deterrence advocacy.[107]

In the postwar era, Szilard used his established technical expertise to promote minimal deterrence as a practical way to mitigate the prospect of nuclear war by meeting arms-race advocates and abolitionists half way. In contrast to the dominant deterrence theories of massive retaliation in the 1950s and mutual assured destruction (MAD) in the 1960s, minimal deterrence held that only a few Bombs were necessary to deter enemy aggression.[108] In the years before the USSR possessed the Bomb, the United States operated according to Secretary of State John Foster Dulles's theory of massive retaliation, by which the United States hoped to deter Soviet aggression by promising to annihilate the USSR with the unmatched magnitude of the United States' stockpile.[109] Dulles mobilized the theory of massive retaliation by evoking the evangelical-moral worldview of a specific, but influential, segment of the American populace as a means for the United States to preserve its geopolitical dominance over the USSR, if not the entire world.[110] For President Eisenhower, massive retaliation was "punishment . . . presented within a strongly moral structure, as a moral right," in the words of Ned O'Gorman.[111] For Szilard, massive retaliation was "morally unacceptable."[112] MAD was even more repugnant. Rather than promising mass murder, MAD promised both mass suicide and mass murder, since it was based on high-level saturation parity whereby the superpowers possessed equivalently large stockpiles.[113] Massive retaliation and MAD are exemplars of "nukespeak," or hiding the horrific reality of Hiroshima and Nagasaki behind euphemisms and metaphors that "encoded" the "nuclear mindset" of politicians, soldiers, and scientists into a self-reinforcing belief that the Bomb had become both a permanent fixture in arsenals and a boon to humanity.[114]

A more moral deterrent, Szilard proposed, would rely on fewer Bombs and threaten fewer people and places. Szilard's most detailed explanation of his particular version of minimal deterrence, published two months before his death in 1964 in the *Bulletin of the Atomic Scientists*, recommended both that the United States and the USSR reduce their stockpiles to twelve comparable warheads, and that each state's citizens should perform rigorous public inspections of their own and each other's weapons programs.[115] This strategy would force nuclear states to "accept the principle of 'one-for-one,'" by which Szilard meant that each state that nuked an enemy city or cities must "tolerate" a comparable level of destruction of one or more of their own cities in return.[116] The United States and the USSR would have to "reach a meeting of the minds," Szilard wrote, and reduce "their strategic striking forces, step by step to a level *just sufficient* to inflict 'unacceptable' damage in a counterblow in case of a strategic strike directed against their territory."[117] Rather than calling for massive nuclear escalation and proliferation, Szilard asserted that restraint and stockpiling only a few Bombs would deter nuclear war, at least for a while. Even with only a few Bombs in existence, their presence and the fear they motivated would mandate restraint and caution.

Minimal deterrence contributed to Szilard's rhetorical being by enveloping his anti-Bomb idealism with a more practical plan for confronting Bomb-struck politicians and soldiers, like Gen. Curtis LeMay, the US Air Force chief of staff who advocated deterrence based on overkill (the maintenance of a nuclear stockpile capable of killing the enemy's population many times over), the increase of US aerial power, and the nuking of Cuba during the missile crisis. While both the United States and the USSR increased their stockpiles, Szilard saw minimal deterrence as a temporary fix in lieu of a permanent solution. It would be, "like the test ban," an incremental step toward disarmament.[118] In his introduction to Szilard's *The Voice of the Dolphins*, atomic historian Barton J. Bernstein described this temporally bifurcated dimension of Szilard's advocacy. "Driven by the fear of mass nuclear destruction, he worked as a political thinker and genial agitator at two levels: formulating long-term strategies to eliminate the bomb and create world peace; and devising short-term tactics to get through a period of crisis and avoid war," Bernstein wrote.[119] Szilard's begrudging advocacy of minimal deterrence became more prominent in his writings as prospects for abolition declined in the 1950s and early 1960s. Szilard argued, "It is probably true that we cannot have general disarmament without also having a far-reaching political settlement. The conclusion of an agreement providing for arms control based on the concept of the minimal deterrent need not, however, await political settlement in Europe or elsewhere."[120] A median point between abolishment and overkill, the strategy still relied on using the weapons he wanted to abolish, which broke his moral concept of a Bomb-free world. But, if adopted, minimal deterrence would halt nuclear proliferation and escalation. Furthermore, minimal deterrence did not make later arms control impossible, and in contrast to the careening arms race, reducing the size of nuclear stockpiles did appear like a significant step toward eliminating them in the long term. With the Manhattan Project long finished, and notwithstanding his atomic realpolitik, Szilard possessed only limited political influence. His authority as a physicist, even when he united with his colleagues in the Federation of Atomic Scientists, and his tenacity met much resistance in Washington, DC, and his realpolitik attitude did not compel any sort of idealized morality in his behind-the-scenes political negotiations.

So when he found that his scientific authority on atomic facts and his anti-Bomb public morality were inadequate to sway politicians at the highest level to adopt minimal deterrence, Szilard turned to atomic fiction to negotiate the Paradox that was presented by living with the Bomb. When "How to Live with the Bomb and Survive," his 1960 article that outlined his basic approach to minimal deterrence, met with resistance, he just "added dolphins and wrote it as fiction" to make policy-makers "face the facts," in the words of one Szilard biographer.[121] His fictional iterations of minimal-deterrence theory appeared as "The Voice of the Dolphins" in 1960 and "The Mined Cities" in 1961. And once again, Szilard presented the destructive and preservative capacities of nuclear deterrence that the Bomb promised. Confronted with the fact of probable, if imagined, global atomic holocaust, Szilard asserted via fact-based fiction that mining each other's cities with Bombs would enable an automatically minimum level of retaliatory power in the event of war, and guarantee the capacity of civilian oversight. Mined cities would provide retaliatory insurance, and the fear of a small number of emplaced Bombs in American and Soviet cities would dictate rational rather than rash military decisions. When Szilard moved from policy writing to fiction, this transition "from a factual to a fictional writer" was not a challenge to facticity but another endorsement of his own version of them, his version of the plan that would promote arms reduction, if not abolishment, when learning how to live with the Bomb's presence.[122] Szilard, however, had no better luck advocating minimal deterrence in fiction than in nonfiction, a result he predicted within the story itself, lamenting in "The Mined Cities" that "the proposal was presented in the form of fiction and it was not taken seriously."[123]

Szilard's fiction thus presented a factual—or, rather, his version of the factual—depiction of what the military future held for the world. His imaginary war ended in "The Mined Cities" with a stable and preservative, if horrifying, peace in the presence of the Bomb. "Presence, and efforts to increase the feeling of presence, must . . . not be confused with fidelity to reality," noted Perelman and Olbrechts-Tyteca,[124] and Szilard's version of atomic Cold War reality, which still relied on the presence of the Bomb, was just as terrifying as massive retaliation and MAD. Science fiction, wrote Thomas Disch, gives people "our basic sense of what is real and what isn't."[125] As a prime example of the circular relationship between fiction and weapons science, Szilard made atomic annihilation seem real, but he did not make minimal deterrence seem like a feasible policy option.[126] Moreover, Szilard's version of the facts, like the rhetoric of MAD, kept humanity under the Bomb's spectral mushroom clouds, clouds that intertwined imaginary atomic blasts with those of distant atomic-weapons tests at far-flung proving grounds and the memory of World War II's conclusion. Szilard compelled his audiences to imagine a horrifying prospect—the guaranteed annihilation of their homes as a means to keep a delicate, preservative peace. In "How to Live with the Bomb and Survive" Szilard had mused that "in Europe, perhaps more than anywhere else, people might rebel at the thought that their city might be sacrificed on the altar of more or less irrational national goals."[127] With his failure to promote minimal deterrence, he found that people the world over rebelled at the thought that their city would be obliterated when "the nuclear field continuously threatened to breach the borders of home and to consume all those huddled 'inside,'" in the words of Bryan Taylor.[128] Minimal deterrence's preservative capacity did not seem like a feasible reality in juxtaposition to the Bomb's destructive capacity, at least in Szilard's formulation. His rhetoric and minimal deterrence reified the same pattern of how human life had been conducted up to that point, rather than pointing humanity toward a fundamentally different way to dwell with the Bomb.[129] Imagining the potentially permanent presence of enemy Bombs within American cities and imagining the kind of war that presence would entail repulsed politicians and scientists alike.

The copresence of the Bomb and words about the Bomb continued to be vexing. In terms of humanity's interaction with the Bomb, rhetorical scholarship has portrayed a bleak outlook. Rhetors, such as "nuclear freeze" proponents and nuclear-power technocrats, have yet to resolve the complex dilemmas posed by nuclear weapons and energy.[130] With Hiroshima and Nagasaki still making their presences felt, descriptions of sudden, mass extermination in fiction and policy discourse became banal. Szilard's paradoxical rhetorical being encountered mixed success. The presence of his words about the Bomb in nonfiction and fiction kept the Bomb's purported ability to annihilate humanity in front of his readers as scientific fact. In retrospect, though, instead of dissuading his readers with catastrophic scenarios, he offered policy-makers, the military, and sci-fi authors both a compelling weapon and a compelling language with which to argue for the weapon's importance. Thus, as an early atomic rhetor, Szilard's rhetorical being paradoxically helped frame the American atomic strategy that he ended up opposing.

3. Conclusion

While Szilard pursued saving the world and making it livable, too, the grandeur of these idealistic goals entailed working toward them in practical, small, and incremental steps. His rhetorical being advanced scientist by scientist, politician by politician, and policy by policy as he aimed first for nuclear arms production, then control, then reduction, then elimination. The rhetorical tactics that characterized Szilard's rhetorical being—manipulating the normative scientific ethos, embodying a unified scientific voice, and realpolitik bargaining—were calculated according to the practical push and pull of negotiating the probabilities of atomic annihilation and atomic peace. As shown by this chapter's middle-way analysis, Szilard's rhetorical being exuded both a public morality and ever-present paradox as he learned to live with the presence of the Bomb, compiling and rearranging the scientific facts entailed by the Bomb's presence in order to fit differing scientific, political, and fictional exigencies.

Szilard's rhetorical being, I suggest by way of a conclusion, displayed a tactical pragmatism that he exercised in the long and unfinished quest for ideal peace. He thereby operated according to the logic of pragmatic idealism.[131] Szilard prudently pursued manufacturing the Bomb as a pragmatic way to achieve the desirable ideal of world peace, and then he prudently pursued arms control with pragmatic rhetoric to achieve the same ideal, to paraphrase Aristotle and Nicholas Rescher, who both pondered how practical rhetoric and action interact with lofty goals.[132] Szilard held that his prudent actions bettered humanity when he both spurred atomic research and sought to abolish its dangerous results, but he dissociated, to an extent, his ideals from his many attempts at policy intervention to make his arguments appear more realistic. Pragmatic idealism made Szilard's weapons rhetoric sound reasonable as he entered the manipulative world of political rhetoric, where his scientific principles and communicative skills might not find traction. When one rhetorical tactic failed, he pragmatically jettisoned it to pursue another tactic, without jettisoning the ideal goal. Szilard never would have abandoned the ideal of world peace, but he did abandon rhetorical tactics when events warranted doing so. Szilard's pragmatic idealism was thus as paradoxical as his rhetorical being. It empowered him to work toward producing the Bomb and work toward abolishing it.

Within the larger realm of weapons rhetoric, pragmatic idealism offers a way of thinking that keeps the polarized extremes of total extermination and permanent peace in mind without hampering day-to-day affairs. Szilard did not position "perfection against failure with no middle road" as a way "to make sense of the technological world" such that "perfection will never be attained and 'failure' will be the only conclusion left," to borrow the words of Harry Collins and Trevor Pinch's portrayal of how technologies tend to be described in dualistic terms.[133] Rather than perpetuating a false dilemma, pragmatic idealism empowers a weapons rhetor with a potential way to avoid touting ultimate destruction and preservation, even when those polarized concepts remain intractable.

Of course, not all atomic rhetors displayed Szilard's penchant for practical idealism. Nonetheless, later Bomb advocates, such as President Eisenhower in his "Atoms for Peace" speech, could appropriate well-established paradoxical arguments for retaining and increasing a nuclear arsenal in the name of peace.[134] And if Szilard's quest for peace represented a minority opinion during World War II, then that stance shifted as the Cold War ensued, and it became official, if not actual, policy to use the weapon for peace rather than destruction. Early in the Cold War, atoms were for peace, not war. But Szilard desired to achieve peace through abolition and reduction, while Eisenhower sent the United States on a "mission for world peace" that entailed an amplified nuclear threat of massive retaliation, continued weapons escalation, and a false appeal to disarmament hopes.[135] Although Eisenhower's "Atoms for Peace" speech included explicit references to disarmament, his appeal to atomic peace functioned as a psychological ruse that dissociated the peaceful President Eisenhower from General Eisenhower, the warrior.[136] Szilard had anticipated such an approach to Cold War politics in his testimony to the House Military Affairs Committee in October 1945 regarding the May-Johnson bill. Szilard, displaying an "unexpectedly practical approach," stated that it was impossible to have an intelligent discussion of atomic technology, when "some people think its function is to build power plants in remote places and others think it is to provide us with atomic bombs so that we can blast the hell out of Russia before Russia blasts the hell out of us."[137] In the end, Eisenhower's policy further entrenched the ideology of deterrence and reiterated Technē's Paradox anew: atomic peace meant enduring the permanent threat of atomic annihilation. Thus, the same paradoxical rhetoric of mass disaster mitigated the ideal of peace in both Eisenhower's and Szilard's atomic policies

The weapon seemed to obtain more influence as a symbol than its actual limited deployment as a weapon would seem to permit, and atomic sci-fi framed the Bomb's continued development and proliferation. As a ubiquitous symbol, novelists and politicians could rhetorically drop the bomb as a commonplace of Cold War politics. The ultimate weapon did not produce peace through deterrence, and it shares the power to motivate people's beliefs and actions with weapons that have much more presence in most people's everyday lives. Remote missile silos, submarines, and even more distant atomic proving grounds remain ever present, but they have become more and more withdrawn, owing to the increasing temporal distance from the two times when the United States dropped the Bomb. Raymond Aron even argued that the Bomb did not cause any substantive change in international relations.[138] It was just another weapon, a more powerful weapon to be sure, but just a weapon. During the Cold War, states operated much as they had before Hiroshima, conducting their wars by nonnuclear means. As such, the capacity of the bomb to wage destruction, win wars, or prevent wars, and even its status as a monolith of foreign relations caused some to overstate its political significance. The Bomb resembles most weapons in their purpose, if not their magnitude of destruction. Thus, the facts of the Bomb and all other weapons can be rearranged to tell a variety of paradoxical narratives about humanity's future and its past.

5

Industrial Antipathy | Irreparability and Ted Kaczynski's IEDs

If there were no computer scientists there would be no progress in computer science.
—Ted Kaczynski (1993)

There's a little bit of the Unabomber in most of us.
—Robert Wright (2001)

On June 24, 1993, Yale University computer scientist David Gelernter picked up a newly delivered package, thinking a graduate student had sent him a dissertation.[1] When he pulled it open, an explosion ripped through his office. Deafened, mangled, and bloodied, he staggered across the street to the campus medical center. The staff looked at him in horror when he stumbled through the doors. He almost died.[2] Ted Kaczynski blew up Gelernter because he was a prominent computer scientist who, in the 1980s and 1990s, had helped to brainstorm and develop internet architecture that supports basic functions such as file sharing, e-commerce, cloud computing, networks, social media, and big data. Kaczynski later sent him a letter—the only time that Kaczynski contacted one of his victims after attempting to murder them—and told Gelernter why he blew him up. Kaczynski called Gelernter a "techno-nerd" whose work contributed to the general destruction of humanity by computerizing the invasion of privacy, empowering computer-facilitated genetic engineering, and generally contributing to environmental devastation.[3] "If there were no computer scientists there would be no progress in computer science," Kaczynski explained.[4] For Kaczynski, Gelernter's work was symptomatic of technology's total and dominating encroachment on human freedom, dignity, and life. Kaczynski saw only one way to avert the coming disaster that these destructive technological encroachments portended for humanity—"a revolution against the industrial system," as he put it at the beginning of his antitechnology treatise *Industrial Society and Its Future*, also known as *The Unabomber Manifesto*.[5]

I argue that Kaczynski negotiated Technē's Paradox by insisting that technology is categorically a force of annihilation. In order to control the rhetorical instability of weapons rhetoric—as Fries and West did, which was the focus of chapter 3—Kaczynski asserted only one pole of Technē's Paradox. Unlike Fries and West, who attempted to stabilize chemical-weapons discourse by categorically defending the weapons as preservative for humanity, Kaczynski categorically rejected the idea that any artifact of the "technoindustrial system" would prove preservative.[6] He insisted that all "organization-dependent technology" is destructive, and that technology will annihilate humanity, or at least what it means to be human.[7] "Industrial-technological society cannot be reformed," and "technology is such a powerful social force . . . it can never be reversed," he wrote.[8] He was adamant and "refused to compromise" regarding these points.[9] To counter this domineering technological power, Kaczynski's rhetoric was as hard as his antitechnological stance, his industrial antipathy, and his bombs.[10] In his manifesto and throughout his essays and letters, Kaczynski used a hard rhetoric. By hard rhetoric, I mean language that does not waver from its certain and ruthless attack on the opposition and anyone who does not fight against it. It tolerates neither reform nor counterarguments. Running throughout his hard rhetoric was a dilemma by which Kaczynski forced his audience to make an irreparable choice between acquiescing to a future of technological disaster or revolting against technology. Perelman and Olbrechts-Tyteca called this type of dilemma "the locus of the irreparable."[11] Also central to this chapter is a middle-way argument in which I claim that Kaczynski's improvised explosive devices spoke for him as much as he used them to persuade. His bombs also spoke with hard rhetoric, challenging Kaczynski with an unyielding counterforce. Kaczynski hoped his IEDs would catalyze the destruction of technology. Instead, these eloquent objects ended up spurring the popular rejection of his antitechnology revolution.

Kaczynski's negotiation of Technē's Paradox differs from that of Malthus, Spies, Fries and West, and Szilard, and his thoughts about technological destruction and preservation deserve a brief overview. First, for Kaczynski, technological destruction refers to the oppressive aspects of technology that threaten humanity with losses of dignity and freedom rather than loss of life. Technology destroys what it means to be human. Although he sometimes addressed the topic of specific weapons, Kaczynski tended to direct his antitechnological attitude toward the whole of technology, including machines, systems, and techniques.[12] I claimed in this book's introduction that all technologies can be associated with Technē's Paradox, and Kaczynski mobilized this generalization by indicting all technologies with the charge that they contribute to the collective power of technology to threaten humanity with its ultimate annihilation. All technologies are thoroughly endowed with their destructive capacity. It is, for Kaczynski, their defining characteristic. Hence, one key assessment of his weapons rhetoric is to understand weapons, including his IEDs, as a synecdoche for technology and vice versa, where the parts represent the whole and the whole represents the parts. As technologies, all weapons are categorically forces of destruction, but the power to kill is only one of technology's destructive capacities. His IEDs were thus intended to be the ultimate destructive devices in their potential capacity to catalyze a revolution that would rid the world of all technologies and, by default, rid the world of most of its human population in the ensuing big collapse. In the end, he thus recommends a solution that is nontechnological—getting rid of all technologies, including weapons such as his own.

Second, technological preservation for Kaczynski refers to the preservation of human dignity and freedom, but not human lives. Securing humanity's freedom from technological domination will preserve what it means to be human—the "freedom and dignity" of "WILD nature" and primitive life.[13] For Kaczynski, the only way technology preserves humanity is through its complete and utter absence, its postrevolutionary disappearance upon civilization's total downfall.[14] Kaczynski thus displays little regard for human life, but high regard for how life is lived sans technology. What path led him to think about preserving humanity by destroying technology and thereby killing off almost everyone?

After four years at Harvard University, Kaczynski spent his postundergraduate years pursuing an academic career as a mathematician, and he completed a doctorate in 1967 with a dissertation on boundary functions at the University of Michigan.[15] The University of California, Berkeley Mathematics Department then gave him a job but, disillusioned, he resigned from his position in 1969. Kaczynski, who later went by the nom-de-plumes FC and Freedom Club, bought a plot of land in rural Montana with the help of his brother, David. He constructed his now infamous cabin and led a simple, meager life. Then, in May 1978, he began his bombing campaign. Kaczynski's IEDs were "embarrassingly ineffectual" gunpowder bombs at the beginning, then pipe bombs, and at the end more elaborate devices armed with homemade blasting caps and an explosive similar to C-4.[16] These bombs had rhetorical intent. "In order to get our [Freedom Club's] message before the public with some chance of making a lasting impression, we've had to kill people," Kaczynski wrote in *Industrial Society and Its Future*.[17] FC's April 1995 letter to the *New York Times* made the rhetorical goals of these devices explicit. "Through our bombings we hope to promote social instability in industrial society, propagate anti-industrial ideas and give encouragement to those who hate the industrial system," Kaczynski wrote.[18] In short, Kaczynski meant his bombs to foment antitechnological revolution. To accomplish this task, he meant his IEDs to "hit" the technoindustrial system "where it hurts," and "to hold people's interest . . . to show them that things are happening—significant things."[19] From loggers to geneticists, Kaczynski targeted seemingly random people whose livelihoods in some way supported or participated in the system he despised. His bombs exploded at advertising agencies and in computer-store parking lots. They disrupted the business of airlines and universities. Kaczynski never hesitated to call himself a terrorist. "This is a message from the terrorist group FC," began Kaczynski's *New York Times* letter, which demanded publication of *Industrial Society and Its Future*.[20] "Over the years we have given as much attention to the development of our ideas as to the development of bombs, and we now have something serious to say," FC continued.[21] His IED attacks were thus propaganda by deed. Through his bombing campaign, which lasted through April 1995 and ended with his arrest in April 1996, he ended up murdering three and injuring twenty-three. He remains imprisoned in Florence, Colorado.

As a murderous antitechnology ideologue who was diagnosed with paranoid schizophrenia, Kaczynski was stigmatized as a lunatic, which made it unlikely that the public would have developed a favorable judgment of his rhetorical capabilities.[22] Jenell Johnson argued that the stigmatization that accompanies a diagnosis of mental illness results in the complete undermining of that person's rhetorical ethos for many, if not most, audiences. Stigmatization is both "an act of rhetorical foreclosure" and a "mark of character" that speaks "louder, and more persuasively, than words ever could" to silence the mentally ill.[23] The stigma of mental disability thereby becomes the lens through which audiences understand the stigmatized person's rhetoric, resulting in their words being dismissed as unreasonable, irrational, unreliable, and "arhetorical."[24] But as Johnson pointed out, being "rhetorically disabled" by a diagnosis and an audience is not the same as being "rhetorically unable."[25] When people are diagnosed as mentally disabled, they should not be denied what Catherine Prendergast called "rhetoricability"—being identified as an able rhetor—even if their actions and words alienate audiences.[26] Therefore, I suggest that granting rhetoricability to Kaczynski is vital to understanding the importance of his weapons rhetoric and his distinctive attempt to challenge the logic of Technē's Paradox.

To examine Kaczynski's rhetoricability as he negotiated Technē's Paradox, I show in section 1 how the theory of rhetorical irreparability undergirds three tactics that characterize the hard rhetoric that he used to assert that technology is always destructive and never preservative. First, Kaczynski diminished all nontechnological problems, eschewing them as mere distractions. Second, Kaczynski diminished the advantages of technologies. Third, Kaczynski amplified the destructiveness of technoindustrial society. In the following section, I describe how Kaczynski's IEDs, no longer in his control, materialized a basic facet of terrorism—the globalized presence of rhetorical violence. In this chapter's middle-way analysis of what it means to be in the presence of both a weapon and words about a weapon, I suggest that his bombs spoke in Kaczynski's stead, motivating beliefs and actions that he did not authorize. I conclude by considering how his repugnant notoriety and his IEDs are remarkable in the longitudinal history of weapons rhetoric.

1. Kaczynski's Hard Rhetoric

Perelman and Olbrechts-Tyteca's "locus of the irreparable" connects the tactics that characterize the hard rhetoric that Kaczynski used to negotiate Technē's Paradox. The locus of the irreparable refers to the rhetorical moment and place when and where a rhetor presents an ultimate choice to a precarious audience for the purpose of dealing with an urgent and dire situation. In its focus on the audience's precarious position and the finality of the choice it must make, the locus of the irreparable can overwhelm the audience.[27] Two basic "values" of the irreparable are the certitude of being in a precarious condition, and the terror of being in an unrepeatable situation. Perelman and Olbrechts-Tyteca wrote that an irreparable decision confronts an audience with the understanding that the future is an infinitude wherein the effects of their decision will last forever.[28] Moreover, "whether the results . . . be good or evil, the irreparable event is a source of terror for man; to be irreparable, an action must be taken which cannot be repeated: it acquires a value by the very fact of being considered under this aspect."[29] Thus, rendering the proper decision at a moment when the certainty of one's future gets decided forever becomes both necessary and pressing.[30]

According to Kaczynski, everyone is poised at a precarious, unrepeatable moment of irreparable decision. Humanity is faced with the certitude of approaching bad times, either through civilization-wide collapse or total technological domination, a unique type of urgent, terrorizing certitude. So, he asked, will individuals let the technoindustrial system become so strong that it will oppress, enslave, or exterminate humanity, or will they revolt against the system, annihilate it, and control their own destiny? The results are final. People cannot revoke their decisions. They will either be tools of the machine or revolutionaries. If they choose to acquiesce to technology, technology will decide their fate. If they choose revolution against the technoindustrial system, they will be free—free but only if they survive the ensuing collapse. To drive his point home, to make humanity understand the stakes of its irreparable technological moment, and to justify his violent acts, Kaczynski diminished all nontechnological problems and the advantages of technology, and amplified the destructive power of technology to insist that technology is, indisputably, a force of annihilation. In these ways, Kaczynski mobilized a generalized and one-sided form of Technē's Paradox whereby any and all technologies possess the power to destroy, while his IEDs synthesized this technological force within a specific type of weapon.

Kaczynski's diminishment of all nontechnological problems is the first tactic that characterizes his hard rhetoric. He made the unique importance and power of technology clear in *Industrial Society and Its Future*. "No social arrangements, whether laws, institutions, customs or ethical codes, can provide permanent protection against technology," he wrote.[31] With technology as the dominant social force, any attempt at altering social arrangements, reforming laws, reorganizing institutions, developing new customs and ethics, and reforming technology would fail to address the central problem of dangerous technology. Therefore, by nontechnological problems, I mean problems that are conceived of as being caused first and foremost by society, politics, tradition, culture, and law rather than by technology. "Winning" the "battle" against technology, Kaczynski wrote, "will require our utmost exertions. We can't afford to stretch ourselves too thin by concerning ourselves with other goals. Instead, we must make the destruction of the industrial system the single overriding objective toward which all our efforts are directed."[32] By dismissing all nontechnological problems as secondary or subordinate to the problem of technology, Kaczynski asked his readers to forgo attempting to overcome particular difficulties for the sake of all humanity. To Kaczynski, humanity was at the tipping point toward technological doom for the first time in its known history, so the moment of irreparability was at hand. According to Perelman, when the locus of the irreparable is at hand, choosing what do next becomes profoundly significant. "It is . . . the unique character of the act which gives it its tragic importance," he wrote.[33] Facing the tragic importance of the irreparable technological moment, Kaczynski refused to admit that any actions—from attacks on his writing style to all leftist political endeavors—had any merit in comparison to the larger antitechnology agenda.[34] By demanding a response to his central claim before countenancing any other arguments about technology, and by including expansive preemptive counterargumentation in his works, he intimated that all other societal problems are red herrings that obscure the primary problem of out-of-control technology. He thus diminished nontechnological problems as mere distractions. This tactic is observable in Kaczynski's attack on leftism, his critique of democratic flexibility, and his subordination of politics under technology. His hard rhetoric is further elucidated by contrasting it with the soft rhetoric—by which I mean rhetoric that embraces paradox, ambivalence, incongruity, polysemy, and uncertainty, and that relies on ambiguity to establish and maintain uncertainty—used by David Gelernter. Soft rhetoric helps rhetors evade taking a firm position on a controversial matter by emphasizing the ambiguousness of complex situations. The rhetoric of Thomas Malthus, August Spies, and Leo Szilard could also fall under this definition.[35]

Kaczynski attacked leftists because, in his estimation, their collective drive to solve nontechnological problems makes them incapable of overthrowing technology. Therefore, for Kaczynski, social justice writ large was a distraction. For him, leftists both thrived in the technoindustrial system and helped to make it stronger by fighting for and achieving political reforms. Leftism, according to Kaczynski, meant fighting against racism, sexism, poverty, sweatshops, neocolonialism, capitalism, imperialism, globalization, genocide, and discrimination in all forms and fighting for gay rights, indigenous people's rights, ethnic minorities, animal rights, the environment, workers, immigrants, "sex education and other psychologically 'enlightened' educational methods," social planning, affirmative action, multiculturalism, and "'social justice' in general."[36] Leftists, he wrote, are "a collection of related types" who "insist that everything is culturally relative," make their drive to enforce tolerance and acceptance a "totalitarian tendency," and include socialists, collectivists, "'politically correct' types, feminists, and disability activists."[37] Needless to say, this critique of leftism alienates much of Kaczynski's potential audience, especially academics.[38] He expressed his desire for everyone to consider how the existence of precarious populations, environmental problems, and injustice makes leftists useful to the technoindustrial system because fighting for people, the environment, and justice releases people's rebellious pressure valve and distracts people from the central technological threat. Racism served as one of his examples. He asked, "Why should we work to give black people an equal opportunity to become corporation executives or scientists when we want a world in which there will be no corporation executives or scientists?"[39] Kaczynski denied neither the presence of racism nor its bad consequences, and he lauded the ideals of social justice and equality. More important to him than the threat of racism, though, was the double threat posed by the leftist tendency to fight racism from within the technoindustrial system. "The left today serves as a kind of fire extinguisher that douses and quenches any nascent revolutionary movement," while the "realization" of leftist goals "would even make the technoindustrial system function more efficiently," he argued.[40] By empowering so many injustices and allowing for their potential elimination, in Kaczynski's estimation, technology divides humanity into an array of fighting populations that keeps them distracted from the power of technology.[41] Therefore, "revolutionaries must somehow circumvent or negate these diversionary tactics," he wrote.[42] Kaczynski thus asserted that all of the nontechnological problems that motivate leftist activism should be dealt with only after first destroying the technoindustrial system.

Kaczynski's lack of interest in promoting feminism is another example of his attack on leftism's diminishment of the technological threat. His antipathy to leftism explains, in part, why he is "uninterested in considering the condition of women in society," in the words of anarchist John Zerzan's assessment of Kaczynski's attitude.[43] He lacked interest in anyone's condition except as it related to his or her domination by technology. Yet it was here, where he took a controversial stance that diminishes the importance of social justice, that his hard rhetoric confronted his readers with the locus of the irreparable. Kaczynski indicated that he wanted all of us, including feminist activists, to stop what we were doing and to start fighting against technology. In the words of Perelman and Olbrechts-Tyteca, Kaczynski was setting "the uniqueness of truth" about humanity's greatest threat "against the diversity of opinion" about what course of action to take.[44] Kaczynski, by mobilizing the irreparable locus, named humanity's seemingly perpetual and universal domination by technology as the unique truth to which humanity must react. In so doing, he also attempted to head off the many opinions about how to react to technological domination that might intersect with the oppression of race, class, and gender. Qualifying the "unique" names who or what has a "precarious existence," Perelman and Olbrechts-Tyteca explained.[45] By qualifying everyone's existence as precarious because of the unique threat of technology, Kaczynski demanded revolution against technology rather than systemic reforms aimed at eradicating the oppression of any particular demographic population. Fighting against the oppression of women, for Kaczynski, meant acquiescing to a system in which technology would continue to oppress women, even if major feminist social justice reforms were enacted. Kaczynski therefore "neglects" the oppression of people when a diversity of nontechnological reforms can ease or eliminate oppression, because eliminating the oppression of some would strengthen "The System's" oppression of everyone.[46] He therefore used harsh words to describe leftists in an attempt to desensitize "hypersensitivity" to the language that people use to describe nontechnological human oppression.[47]

Kaczynski's attack on leftism is a more specific instance of his broader dissatisfaction with the flexibility of democratic governance, a dissatisfaction that compelled his diminishment of nontechnological political problems. "The technoindustrial system is exceptionally tough due to its so-called 'democratic' structure and its resulting flexibility," Kaczynski wrote in "Hit Where It Hurts."[48] As a matter of process, the normal course of governmental interaction with citizens never results in a solution to the overarching technological problem. "In a 'democratic' system, when social tension and resistance build up dangerously the system backs off enough, it compromises enough, to bring the tensions down to a safe level," he wrote.[49] Kaczynski's critique aligns well with 1960s thinkers who questioned the pluralistic concept of tolerance. Tolerance is a "state of mind," wrote political philosopher Robert Paul Wolff, that makes allowances for social disturbances of all kinds on the local level as a necessary component of realizing democratic pluralism and its promotion of overall stability, upward mobility, and social justice.[50] Yet, Wolff wrote, "pluralism is fatally blind to the evils which afflict the entire body politic, and as a theory of society it obstructs consideration of precisely the sorts of thoroughgoing social revisions which may be needed to remedy those evils."[51] This is a question of how to balance working toward "the common good" versus eliminating "distributive injustice."[52] In the words of Herbert Marcuse, "it is the whole which determines the truth . . . in the sense that its structure and function determine every particular condition and relation. Thus, within a repressive society, even progressive movements threaten to turn their opposite to the degree to which they accept the rules of the game."[53] Kaczynski described technology as the whole, the truth, the structure, and the function, so any action that used political means to solve problems among citizens only served to use "the parts" to strengthen "the whole." The more a democratic plurality could weed out violence and oppression, the stronger technology would grow, because, in this critique of democratic flexibility, or tolerance, there would be less reason for people to fight against technology if a democratic society tolerates a diverse range of people. For Kaczynski, freedom within a democratic system of governance would become more disastrous with respect to the power of technology to absorb political change. According to Kaczynski's logic, as populations appear to attain freedom and dignity within a technological system with successful social justice movements, technology, when unchallenged, would continue to rob all people of their freedom and dignity.

Nuclear terrorism exemplifies how democratic flexibility makes the technoindustrial system more powerful. One might think that politics and war—"the continuation of policy by other means," as Clausewitz put it—would be the proper arenas in which to deal with the threat of nuclear terrorism.[54] But for Kaczynski, the political focus on terrorists and terrorism was yet another distraction from the central problem of technology. "If Al Qaeda should set off a nuclear bomb in Washington, DC, people's reaction will be, 'Get those terrorists!' They will forget that the bomb could not have existed without the previous development of nuclear technology," he wrote.[55] Just as Eisenhower's "Atoms for Peace" applied a reversed logic that put the effects (peaceful atomic energy use) before the cause (the quest for the Bomb), so, too, does the fight against terrorism put the effects of terroristic violence (the security state) before its cause (the development of the weapons used by terrorists).[56] As long as the United States' populace tolerates its government's solutions to the problems of terrorism, the fight against terrorists will strengthen the technoindustrial system in the long term, Kaczynski surmised, regardless of the destabilization caused by terrorism in the first place. Meanwhile, the technological power that supports both the United States and terrorist organizations would go unquestioned and unchecked.[57] "That's how it is with the 'democratic' industrial system: It gives way before protest, just enough so that the protest loses its force and momentum. Then the system bounces back," Kaczynski wrote from prison, reflecting on his own terroristic violence.[58] A terrorist nuclear blast would thus be bad for victims, but great for expanding the US security state and the technoindustrial system as a whole. In their failure to spark revolution and murderous criminality, Kaczynski's IEDs were also culpable for helping to legitimate the expansion of antiterrorist security.

Kaczynski further diminished political problems by insisting that his revolutionary agenda was apolitical and would destroy the government only as an ancillary effect of destroying the technoindustrial system. For Kaczynski, technology could dominate humanity irrespective of governmental structure, so politics was, without exception, less important than and subordinate to technology. FC's 1995 letter to the *New York Times* stated Kaczynski's loose ties to anarchism, and *Industrial Society and Its Future* confirmed that Kaczynski's revolution was not, foremostly, political. Rather, "the exclusive target of revolution must be technology itself," he asserted.[59] Kaczynski was an anarchist by default, since the level of destruction entailed by the complete demise of technology would include the destruction of all types of government, or political technologies, in the quest for primitive life. Kaczynski scholar David Skrbina called this semiapolitical philosophy "technological anarchism."[60] Thus, Kaczynski saw any actions against government, religion, and capitalism in the name of anarchy as yet more distractions from the central goal of taking down the technoindustrial system, but these anarchist nemeses would be eliminated along with technology.

Contrasting Kaczynski's withering critiques of leftism, democratic flexibility, and political action with David Gelernter's soft rhetoric helps to contextualize why Kaczynski brooked no waffling and wavering on the central technological problem. Gelernter's soft rhetoric wavered about some of technology's commonplace philosophical dilemmas, such as the neutrality of artifacts, determinism, and the accountability for disaster, all while displaying an ambivalent attitude toward the inseparability of technology's positive and negative effects. Gelernter believed that he was Kaczynski's ally, owing to his confrontation with and questioning of technology's negative attributes.[61] But from Kaczynski's viewpoint, the malleability of Gelernter's soft rhetoric proliferated unresolved counterarguments that, in effect, script behavior by absorbing humanity into a life of technological domination and capitalist hierarchy.

Gelernter's refusal to reject technology categorically, after questioning its pros and cons, and his continued work inventing a computer-based human reality that displays a flippant disregard for technology's invasion into personal privacy showed how Gelernter and Kaczynski were in no way conceptual allies. Gelernter described his "mirror worlds" as "software models of some chunk of our reality" in which enormous amounts of data "mimic *the reality's* every move, moment by moment" or "a huge institution's moving, true-to-life mirror image trapped inside a computer."[62] The proposed software would visualize the infinite movements of complex physical systems and institutions, from hospitals to transportation grids, and from banks to social networks, and this visualization would facilitate computer navigation of these systems. Of course, to navigate a mirror world, a user must enter into it and succumb to the program, and the software would have to know every iota of available data about that user in order for its virtual banking, healthcare, transportation, communication, political, work, leisure, and education platforms to function. Gelernter embraced all of this as acceptable despite the obvious drawbacks. His ambivalence about positive and negative technological effects, for instance, is exposed in his oft-repeated assertion of the commonplace argument in technological discourse that new technologies are "inevitable."[63] In this way, Gelernter's *Mirror Worlds* exuded the type of ambivalent and neutral attitude toward technology that aggravated Kaczynski.

No, said Kaczynski—with both a word and a bomb—to Gelernter's vision of virtual humanity. "Never forget that it is the technology itself that has to be eliminated," Kaczynski exhorted again and again.[64] Unlike social justice and other political issues, Gelernter's software project was a primary element of the central problem of technology. For Kaczynski, destabilizing and destroying the computer industry with terrorism was not a diversionary action. Faith in his revolutionary program entailed stopping the implementation of mirror worlds by killing their lead designer. As J. Robert Cox wrote of the irreparable with respect to environmental rhetoric, the "sense of precariousness is captured in references to what is (1) fragile and (2) established, stable, or secure, but threatened by radical intrusion. That which is fragile requires protection or an agent's active intervention to ensure its continued existence."[65] What it means to be human—to live in Kaczynski's wild nature—is threatened by the "radical intrusion" of false reality. Human fragility requires our protection. Thus, Kaczynski's diminishment of nontechnological problems helps to lead his readers to an irreparable place where they must choose to revolt against technology or be forced to live within the mirror world.

Concomitant to minimizing the importance of nontechnological problems by portraying them as diversions, Kaczynski also used the figure of diminishment to refute and downplay any and all beneficial aspects of technology. By refuting the idea that technologies benefit humanity, Kaczynski thereby refuted the idea that weapons or any other technology can be preservative. Unless, that is, a technology has the power to destroy all of technology and, hence, itself. Kaczynski's hard rhetoric diminished technological advantages by contrasting them with the overall disadvantage of technology, subverting the difference between good and bad technologies, and attacking the idea that scientists work on new technologies to benefit humanity.

Kaczynski recognized that refuting the idea that technology benefits humanity is a difficult argument to make, especially in a society engulfed by technology. "Electricity, indoor plumbing, rapid long-distance communication. . . . How could one argue against any of these things, or against any other of the innumerable technical advances that have made modern society?" he asked in *Industrial Society and Its Future*.[66] Kaczynski did not attempt to argue against specific beneficial technologies, such as medical treatments for cancer patients. Rather, he answered his question by emphasizing that technology should be judged as an entire system and not according to individual technologies. If one judges individual technologies, for which "each step" in innovation "will be equally humanitarian in its goals," as he put it in an untitled essay, then the vast majority of technologies can be portrayed as beneficial.[67] Judged as a whole, however, technology is a detriment. Kaczynski admitted that framing a multitude of benefits as a singular massive disadvantage is a "paradoxical notion."[68] However, he asserted, technology provides a simple formula for understanding this paradox. "The system makes an individual's life easier for him in innumerable ways, but in doing so it deprives him of control over his own fate," he wrote.[69] In a letter to Skrbina, Kaczynski explained how curing mental disorders and saving the environment exemplified this paradox: "Though improbable, it's conceivable that the system might some day succeed in eliminating most mental disorders, cleaning up the environment, and solving all its other problems. But the human individual, however well the system may take care of him, will be powerless and dependent. In fact, the better the system takes care of him, the more dependent he will be. He will have been reduced to the status of domestic animal."[70] Kaczynski further explained that the supposed increase in humanity's standard of living, which would result from these beneficial technological advances, was "the system" patting itself on the back for succeeding according to "the system's" own self-determined standards.[71] For Kaczynski, living standards were measurable by access to nature, not electronic gadgets, psychiatry, medicine, communication, plumbing, and all other technological conveniences.[72] Short-term technological successes would lead to "long-term demise."[73] Thus, for Kaczynski, the advantages of individual technologies are not as important, even in sum, as the massive disadvantage of technology as a whole.

In terms of diminishing the benefits of a particular weapon, the Bomb was exemplary for Kaczynski, who rejected the preservative aspect of deterrence theory. On the surface, "nuclear weapons," which were "perhaps the star exhibit" of the domination of humanity by "large organizations . . . offer no benefits whatever—only death and destruction," he wrote in another letter to Skrbina.[74] Deterrence would not keep the peace, he explained. "With the exception only of a tiny minority of dictators, military men, and politicians who see nuclear weapons as enhancing their own power, virtually every thinking person agrees that the world would be better off without nuclear weapons," he wrote.[75] Despite the nearly universal abhorrence of the Bomb, according to Kaczynski, people were not free to rid themselves and their planet of the threat, much less the geopolitical, networked technological systems that sustain it. The one benefit of nuclear deterrence, for Kaczynski, did not justify technology writ large, much less the existence of nuclear arsenals.

People should not bracket the supposed advantages of a technology such that they do not consider the technology's obvious disadvantages, according to Kaczynski. It should not be surprising that in his iteration of Technē's Paradox in *Industrial Society and Its Future*, Kaczynski, as a Jacques Ellul enthusiast, echoed the third of Ellul's four technological rules, which holds that "pernicious effects are inseparable from favorable effects."[76] In the section of *Industrial Society and Its Future* similarly titled "The 'Bad' Parts of Technology Cannot Be Separated from the 'Good' Parts," he intimated that Technē's Paradox might function as a type of gauge to help the public assess technology.[77] Kaczynski here articulated a type of risk analysis that was biased by predetermined hatred of technology. Medicine, genetic engineering, and sanitation served as exemplars of the overarching "bad" technological effect that was of more importance than many lesser "good" effects. In neo-Malthusian fashion, he argued that while medicine benefits humanity by preserving and extending life, it will do so too well, resulting in a gloomy scenario in which "the only solution will be some sort of eugenics program or extensive genetic engineering of human beings, so that man in the future will no longer be a creation of nature, or of chance, or of God . . . but a manufactured product."[78] Good health and long life thus come with the probability that a program of systematic extermination and engineered docility will arise to temper the population expansion empowered by medicine. "The immense power of biotechnology" likewise presents humanity with "irresistible" and "obviously and unequivocally good" effects that will only serve to further enslave people to the destructive "needs of the industrial-technological system."[79] His assessment of sanitation, by which the advantages of increased health and longevity are offset by the rise of allergies, intestinal disease, and the "population explosion," follow the same logic.[80] Better to check the population now by overthrowing the technoindustrial system with one massive positive check, he averred, than to let a miserable Malthusian scenario play out over the long term. In general, Kaczynski acknowledged that technologies do solve technological problems in limited contexts. Rather, the primary problem for Kaczynski was the new, worse problems that the technological solutions would cause as they further thwarted human freedom and dignity.[81]

Kaczynski further diminished the benefits of technology by refuting the idea that scientists who develop technologies desire to "benefit humanity."[82] Kaczynski identified three reasons why the claim that scientists work for the benefit of humanity is unviable. First, he accused scientists of illegitimate personal motivations, such as curiosity, rather than societal ones.[83] Second, Kaczynski dismissed the entire semiscientific fields of archaeology and comparative linguistics as irrelevant to "the welfare of the human race," while giving faint praise to other fields such as research into vaccines and air pollution that at least provide some form of welfare.[84] As a result, Kaczynski gave more academics a reason to despise him. His third critique of science was more important to his diminishment of technological benefits. Scientists who work for humanity's welfare still "present obviously dangerous possibilities," he wrote.[85] Scientists, Kaczynski asserted, do not arrest their research in order to eliminate or prevent technological danger, even when it is obvious. Edward Teller, who was a central figure in the creation of hydrogen bombs, exemplified Kaczynski's idea of a scientist whose "emotional involvement with nuclear power arose not from a desire to 'benefit humanity' but from the personal fulfillment he got from his work and from seeing it put to practical use."[86] Kaczynski imputed that since the destructive potentials of atomic technologies were obvious, then Teller must have developed them for personal, not humanitarian, reasons. "If he was such a humanitarian then why did he help to develop the H-bomb?" Kaczynski demanded.[87] According to Kaczynski, Teller was not at all humanitarian in his pursuit of the Bomb, regardless of any benefits that might have derived from "cheap electricity."[88] For Kaczynski, Teller was a bad person and so were his inventions. Kaczynski viewed the benefits of atomic energy as a ruse that obfuscated enormous technological harms, and his hard rhetoric brooked no prevaricating about Teller's negative contributions to the technoindustrial system. To choose acquiescence to the technoindustrial system thus entails making the irreparable choice to grant tacit support to hydrogen bombs and their inevitable use.

The amplification of technological destruction was the third main tactic by which Kaczynski's hard rhetoric confronted his readers with the locus of the irreparable. The impetus for his desire to amplify the destructive aspects of technology might have come from Kaczynski's Harvard education, as suggested by his biographer, Alston Chase. Chase indicated that the Harvard curriculum, reformed according to 1945's *General Education in a Free Society*, emphasized the dual poles of Technē's Paradox as a type of central philosophy, or modus operandi, peculiar to twentieth-century life. According to Chase, courses at Harvard played an important role in stimulating Kaczynski's industrial antipathy by inculcating a positivist scientific mindset with a "despairing message" about the negative effects of science and its technologies.[89] The optimism exuded by positivism and the life sciences that science would forever enhance human life for the betterment of humankind competed with the dismal portrayals of technological society by the likes of Karl Marx, Friedrich Nietzsche, Thorstein Veblen, Oswald Spengler, Norbert Wiener, Lewis Mumford, Herbert Marcuse, and most of all, Jacques Ellul.[90] Kaczynski even clarified that he was parroting the antitechnological views he found elsewhere.[91] One example of his education in the Paradox at Harvard is his firsthand experience of both the preservative and destructive aspects of psychological technologies as the subject of unethical experiments by psychologist Henry A. Murray. The study of psychology is taught as a way to better humanity by understanding the mind and brain. But over three years, Murray subjected Kaczynski and other undergrads to a withering barrage of interrogations and other tests meant to undermine their mental health in order to make observations about how people react to adversity and alienation.[92] These experiments seemed more destructive than preservative of the dignity of human life. Kaczynski rejected the positive outlook on technology, accepted the destructive capacity of technology as endemic, and proclaimed the detrimental effects of technology again and again. His amplification of technology's destructive capacity is observable in his demonstrations of our collective loss of humanity, his list of technologies that threaten humanity, the attribution of autonomous technological determinism to technology writ large, and the call to destroy it all as the one and only option.

Kaczynski amplified technological destruction by drawing attention to the many ways that technology destroys what it means to be human. At times, Kaczynski took an exterministic position. "The unrestrained growth of technology," Kaczynski wrote in "The Coming Revolution," was a totally annihilating force that "threatens the very survival of the human race."[93] Kaczynski, however, disbelieved that technological human self-extinction will take place anytime soon. Rather, people face "disaster of another kind" in the immediate and ongoing "loss of our humanity."[94] A purposeful human life, for Kaczynski, was defined by freedom and dignity.[95] Technologies like artificial intelligence and genetic engineering threaten both.[96] Sounding quite patriotic, and summoning the revolutionary rhetoric of France and the United States, he warned in his untitled essay of "the extinction of individual liberty" by technology.[97] In an "irreversible" process within the technoindustrial system, people would put their individual and collective fates into the hands of "politicians, corporation executives and remote, anonymous technicians and bureaucrats whom he as an individual has no power to influence," he wrote in *Industrial Society and Its Future.*[98] To paraphrase Cox's explanation of the irreparable locus, Kaczynski claimed that technology's destructive force entails specific consequences for everyone, for we all must decide whether to revolt against technology while at risk of being complicit in causing the "irreplaceable loss" of our humanity.[99] It might be odd to think of Kaczynski as a humanitarian, but he displayed a deep regard for what makes us human at the same time that he held contempt for technoindustrial human life.[100] Thus, Kaczynski amplified technological destruction because he was confident that humans will become technology's pets if humans fail to revolt.

Kaczynski amplified technological destruction in a second way by listing the sheer number of dangerous technologies that exacerbate the technoindustrial system's encroachments on humanity. In the foreword to *Technological Slavery,* Kaczynski emphasized that one of "the four main points that I've tried to make in my writings" is that "technological progress is carrying us to inevitable disaster."[101] Moreover, "the longer the technoindustrial system continues to expand, the worse will be the eventual disaster."[102] What technologies threaten humanity? His "partial list of problems" is long:

> War (with modern weapons, not comparable to earlier warfare), nuclear weapons, accumulation of nuclear waste, other pollution problems of many different kinds, global warming, ozone depletion, exhaustion of some natural resources, overpopulation and crowding, genetic deteriora-

> tion of humans due to relaxation of natural selection, abnormally high rate of extinction of species, risk of disaster from biotechnological tinkering, possible or probable replacement of humans by intelligent machines, biological engineering of humans (an insult to human dignity), dominance of large organizations and powerlessness of individuals, surveillance technology that makes individuals still more subject to the power of large organizations, propaganda and other manipulative psychological techniques, psychoactive medications, mental problems of modern life, including inter alia, stress, depression, mania, anxiety disorders, attention-deficit disorder, addictive disorders, domestic abuse, and generalized incompetence.[103]

Throughout his writings, genetic engineering, biotechnology, mass media, nuclear energy, and the chemical industry were some of Kaczynski's most frequent targets. Each of these dangerous technologies could prove catastrophic on their own. All together, though, they represented irrefutable proof for Kaczynski that humanity is doomed. In a letter to Skrbina, Kaczynski wrote, "Even if we make the extremely optimistic assumption that any one of the [technological] problems could be solved through reform, it is unrealistic to suppose that *all* of the most important problems can be solved through reform, and solved in time" to prevent disaster.[104] In its level of amplification, the long list of dangerous technologies is overwhelming. Nonetheless, the list demands a response to Kaczynski's central claim that all technologies are destructive, and it provides additional evidence for his subsidiary claim that any responses to his arguments that fail to address the destructive capacities of technologies are tangential digressions that serve to empower the technoindustrial system further.

In terms of this book's middle-way argument about the mixed agency of rhetors and technologies, Kaczynski can be seen as attributing a much more powerful causal agency to technology than he does to either humans or human/nonhuman assemblages. Although Kaczynski believed in the power of humans to revolt against the technoindustrial system, he was a technological determinist through and through. And his promotion of technological determinism was a third way that he amplified technological destruction. "The development of the technoindustrial system cannot be controlled, restrained, or guided, nor can its effects be moderated to any substantial degree," Kaczynski wrote.[105] After all, he was a student of nineteenth- and twentieth-century philosophers of technology who conceived of technology as an all-powerful, dominating force. Citing Marx, Ellul, Samuel Ramos, and Samuel Butler, Kaczynski wrote, "it is technology that rules society, not the other way around."[106] Industrial society was, hence, a "social machine."[107] And when the social machine destroyed, it did so of its own accord and could destroy as much as it willed. Kaczynski was adamant that neither more technology nor reforms of the technoindustrial system would stop the destructive impulses of autonomous technology. It threatened humanity's survival and pushed us to the brink, the locus of the irreparable.

Whereas nineteenth- and twentieth-century philosophies of technology tended to look to capitalistic, communistic, or environmental politics, more technologies, new attitudes, refusal, or God to get out of the technological dilemma, Kaczynski exhorted violent "revolution against the industrial system."[108] The amplified destructive character of technology both warranted and necessitated it. "If we want to defend ourselves against technology, the only action we can take that might prove effective is an effort to precipitate the collapse of technological society," Kaczynski wrote, again announcing the terms of irreparability.[109] Cox wrote that the locus of the irreparable warrants "extraordinary measures," or "actions which go beyond the usual, customary, or what most people would approve."[110] Revolt against technology is such an extraordinary measure. But a problem with invoking the locus of irreparability, as noted by David Zarefsky, is that when delivering the news of a tough dilemma to a "diverse audience," many people will oppose the argument, most "will be uncommitted," and only a few will make the difficult choice to, in this case, revolt.[111] Hence, the moment of advocating revolt is a key moment when Kaczynski's rhetoric is at its hardest. "Only the collapse of modern technological civilization can avert disaster," he wrote as one of his "four main points." And in *Industrial Society and Its Future*, he wrote that "it would be better to dump the whole stinking system and take the consequences."[112] Kaczynski was not diverted from his goal of bringing down technological society, and his rhetoric did not waver. In "The Coming Revolution," Kaczynski wrote that the "scientists, engineers, corporation executives, politicians, and so forth, who consciously and intentionally promote technological progress and economic growth are criminals of the worst kind," even "worse than Stalin or Hitler."[113] He would "hit them where it hurts" and he urged everyone to do likewise in order to ensure that "when industrial society breaks down, its remnants will be smashed beyond repair, so that the system cannot be reconstituted. The factories should be destroyed, technical books burned, etc."[114] Thus, Kaczynski amplified technology's destructive capacity in order to goad humanity's destructive capacity into action.

Taken together, Kaczynski's diminishment of nontechnological problems, diminishment of technological benefits, and amplification of technological destruction demonstrated both how unwavering his hard rhetoric was and how the locus of the irreparable depends upon uniqueness, precariousness, and timeliness.[115] As a manifestation of the generalized form of Technē's Paradox, these rhetorical tactics meshed well with a strategy meant to tap into commonplace anxieties about technologies, and demonstrated that, in the words of Skrbina, we are all "morally required to work to take the system down, by any means necessary."[116] Kaczynski's negotiation of Technē's Paradox as a generalized technological predicament, rather than one bound to a particular weapon, revealed the irrevocable, irreparable action that the precariousness of the overarching technological situation required. Revolutionary violence should be afoot for those who concur that nontechnological problems are subordinate to technological ones, that the advantages conveyed to humanity by technology are overblown, and that technology is too destructive. Yet the advocacy and perpetration of terroristic violence was the point at which Kaczynski lost his audience. He might have tapped into a generally popular technological antipathy, but he had murdered and maimed people with indiscriminate terror.[117] Thus, the moments when his IEDs ripped into unsuspecting victims not only marked the step from condemning the technoindustrial system to blowing it up, but also marked the point at which people judged Kaczynski for his "specific misdeeds" rather than his "general principles."[118] The moments when his IEDs exploded, his hard rhetoric became subordinated to his bombs' own particular rhetorical force as eloquent objects.

2. Rhetorical Violence: Talking Bombs

This chapter's middle-way approach to the rhetorical character of weapons examines the presence of Kaczynski's IEDs as a type of entailed rhetorical violence. As a textual analysis, one might say the preceding section examined Kaczynski as a type of "literary terrorist" who dropped "cultural bombs" and set off "literary explosives."[119] But revolutionaries, terrorists, and states do not only persuade and dissuade the commission of violent acts with spoken and written weapons. They commit violence at the same time they enunciate it.[120] Most of Kaczynski's IEDs, though, exploded long before his words did. The tendrils of his infamy spread along flight paths and postal routes before materializing in an FBI dossier and the mass media. In this materialization, Kaczynski's words and his IEDs are as inseparable as they are discrete entities. Rhetoric and violence in general are also both inseparable and discrete insofar as they both motivate belief and actions. Rhetoric and violence—words and weapons—"are at once both irreducible to each other and inextricably interrelated," to borrow Jean Baudrillard's gloss on the relationship between good and evil.[121] As discussed in chapter 2, August Spies lost control of the meaning of the word "dynamite" but retained the power to control his identity if not his biological fate. In contrast, the vexed interactivity between rhetoric and violence disempowered Kaczynski, who lost control of the meaning of his IEDs as much as he lost control of his identity when the FBI dubbed him "UNABOM."[122] He had particular goals when he constructed each IED, but Kaczynski did not know "what its precise powers would be," although he might have surmised that the ensuing events would adhere to a "graspable pattern," to use the words of Andrew Pickering's description of the "emergence" of material agency.[123] Once Kaczynski remade himself by exploding his IEDs, he found himself remade once again when they were turned back against him.[124] To understand Kaczynski's IEDs, I define their rhetorical violence as the motivation of human thought and belief that can be traced to violent events and to weapons, and that escapes the control of those who perpetrate the violence. The unintended consequences entailed by the presence of weapons sometimes displace the agency of those who wield them. By naming and defining rhetorical violence, I do not intend to equate violence and rhetoric, nor do I intend to separate them. Violence and rhetoric are at odds, and yet they work in tandem as rhetorical violence. Examining the concept of rhetorical violence with respect to Kaczynski's aim to "hit where it hurts," the reactions of the FBI and of the legal apparatus to his IEDs, the confusing interchangeability of humanity and technology as his targets, and the terroristic omnipresence of the bombs demonstrate that Kaczynski's rhetorical agency was displaced by the Unabomber's IEDs.

There are several ways to define rhetorical violence from a material perspective. Violence and weapons are not rhetorical figures or tropes, at least in the classic Greco-Roman sense, although they can be used as such. Violence and weapons possess a motivational force, and it is not for unjustifiable reasons that violence, war, and weapons have served as metaphors for rhetoric, debate, and coercion. The history of rhetorical theory is littered with examples. The mighty figure of Rhetorica appears with her sword drawn. "To be injured by speech," in the words of Judith Butler, happens when someone enacts a violent agency via the force of words, such as with hate speech, which puts "the addressee out of control."[125] Moving from the results of biological injury and death to their causation in persuasive exhortations to maim and kill is another form of rhetorical violence.[126] Human bodies can become "an extension of the weapon" via the language of killing.[127] Then there is symbolic violence, or the mobilization of violent events for persuasive purposes. One might also commit an act of violence with the intent of using the event as a symbolic means to motivate belief and action, which is propaganda by the deed.[128] Kaczynski used the symbolic violence of his own terrorism to, for instance, negotiate the publication of *Industrial Society and Its Future* and ensure that people would have a reason to read it, which is an act of coercion or blackmail. All of these forms of rhetorical violence pertain to weapons rhetoric, and even casual readers of this book will recognize that rhetorical violence has been present throughout its pages. Kaczynski's case, however, provides a peculiar opportunity to consider rhetorical violence with respect to a person who designed, built, and detonated his own weapons in addition to perpetrating a guerilla mass-media campaign to legitimate their use. Unlike the other weapons rhetors in this book, Kaczynski controlled the lifespan of his weapons from idea to explosion. Furthermore, the difference between the presence of words and the presence of words about weapons marks the point at which Kaczynski's terrorism splintered away from the antitechnology philosophy of Jacques Ellul.[129] Once his bombs exploded, when they started to talk, they took on a different type of presence than that which was authorized by Ellul or designed and rationalized by Kaczynski.

The essay "Hit Where It Hurts" shows that Kaczynski understood violence as rhetorical and vice versa. More than just listing "the vital organs of the system" that "radicals must attack" (the electric power industry, the communications industry, the computer industry, the propaganda industry, and the biotechnology industry), the essay focuses on the uses of and occasions for persuasion.[130] He even used the rhetoric-violence metaphor to structure the essay's purpose: "I have to explain that when I talk about 'hitting where it hurts' I am not necessarily referring to physical blows or to any other form of physical violence. For example, in oral debate, 'hitting where it hurts' would mean making the arguments to which your opponent's position is most vulnerable. In a presidential election, 'hitting where it hurts' would mean winning from your opponent the states that have the most electoral votes. Still, in discussing this principle I will use the analogy of physical combat, because it is vivid and clear."[131] Although

Kaczynski might not have been "necessarily referring" to physical violence, he was by implication referring to violent acts. As he noted, he could not "recommend violence of any kind," owing to his incarceration.[132] His caveats about not recommending violence and criminality while attacking technology through "legal means, of course," were ironic.[133] To extend Kaczynski's pugilistic metaphor, "hitting where it hurts" does not refer to knocking out any particular industry or the entire technoindustrial system. Rather, the rhetorical purpose of "hitting where it hurts" is to goad a particular industry or institution into a long, grueling bout that will give the underdog a chance at, in the end, defeating the champion. Owing to the "flexibility" of the "democratic structure" that defuses public worry and anger about industrial society, "in order to hit the system where it hurts, you need to select issues on which the system will *not* back off, on which it will fight to the finish . . . the vital organs of the system," he wrote.[134] In juxtaposition to such blows that draw an opponent into a fight to the death, "trotting off to the next world trade summit to have temper tantrums over globalization" will neither hurt the technoindustrial system nor force it to react.[135] The World Bank can just ignore such protest, and leave skirmishing with the protestors to an overmilitarized security squad. The idea of the World Bank fighting to the finish with protestors who use traditional means of dissent is unthinkable. Instead, winning the key issue of the debate—or, rather, bombing the most vulnerable key personnel—of critical industries will "hurt" the system.[136] Thus, "hitting where it hurts" entails seeking engagement with power and authority in order to elicit an aggressive response that revolutionaries can turn into both a public debate and a violent conflict. But while persuading the key people who sustain the technoindustrial system to desist from their livelihoods is within the realm of possibility, it is more plausible according to Kaczysnki's logic that IED explosions will force the technoindustrial system to react by "fighting to the finish."

Consider the technoindustrial system's reaction to the Unabomber's IEDs as a way to gauge the power of his weapons to instigate the type of "fight to the finish" he imagined. His IEDs functioned beyond the scope of the rhetorical goals he identified, and instead he ended up instigating a fight to his own finish. Kaczynski's blackmailing of the *Washington Post* and the *New York Times* to secure publication of *Industrial Society and Its Future* succeeded, but his additional demands to continue publishing responses to his critics in the same newspapers never materialized. His arrest weeks after the publication of the treatise ended his chance to continue writing for the media from which he had, with the FBI's approval, coerced initial cooperation. In the long run, his terrorism was as effective as throwing antiglobalization tantrums even though he, according to his own definition, "hit where it hurts." Instead of motivating either a pervasive national conversation about technological dangers or a withering debate within the news media, his IEDs told the state and the mass media to stigmatize his antitechnology attitude as insane. His methods hit him where it hurts.

When his bombs spoke for him, they did so as a type of rhetorical violence that spoke more directly to much of his audience than his words did. Arguing that Rorschach-test cards speak as objects, Peter Galison made a cogent statement that bears upon the rhetorical violence of Kaczynski's IEDs. "Just insofar as these cards are described, they describe the describer. Not only do these objects talk back, they immediately double the observer's language with a response that pins the speaker on a psychogrammatic map. These are the cards of the Rorschach test; and they don't mind sending you home, to the clinic, or to prison," Galison wrote.[137] Furthermore, such "objects also make subjects: 'depressive,' 'schizophrenic.'"[138] As much as Kaczynski described his IEDs, they in turn described the describer, with perhaps greater rhetorical effect. When they spoke, they sent him to prison and pigeonholed him with a damaging psychological diagnosis by helping to label him a paranoid schizophrenic. If one believes in the validity or sanity of his rationalizations, then his bombs offered a counterargument, even as they corroborated Kaczynski's thesis that technology is categorically a force of destruction. His IEDs destroyed the Freedom Club. Kaczynski might have surmised that would happen.

Weapons force governments to respond to their presence.[139] In Kaczynski's case, his IEDs helped to determine his legal fate by warranting the state to brand him a paranoid schizophrenic, which forced him to abandon his intent to justify his actions at trial. Upon his arrest, the legal proceedings and Kaczynski's official psychological assessment took many twists and turns. Confronted by the media's insistence that he was a deranged killer, he was adamant that he was not. Court-appointed psychologists at first agreed with Kaczynski, but that assessment was overturned by a second psychological profile created by Sally R. Johnson, who based her diagnosis of Kaczynski's schizophrenia on his antitechnology views, the facts of his bombings, and his belief in his sanity. Xavier F. Amador, an advisor who helped to manufacture Johnson's state-sponsored "mental defect" defense, declared Kaczynski to be suffering from schizophrenia complicated by anosognosia, the belief in one's own sanity and the desire to prove it in the face of "life threatening consequences" that instead constitutes proof of one's insanity.[140] In sum, Johnson's report labeled Kaczynski's industrial antipathy insane, his violence insane, and his belief in his sanity insane.[141] Kaczynski tried to circumvent this official diagnosis by asserting his right to represent himself in court. Michael Mello, one of Kaczynski's legal advisers, noted, however, that allowing Kaczynski to defend himself was tantamount to assisted suicide, because mounting a political defense of his crimes was destined to fail.[142] His court-appointed lawyers' staunch attitude against the death penalty provided a further obstacle for Kaczynski. Led by attorney Quin Denvir, they would do anything in their power to keep Kaczynski alive. Kaczynski was deemed capable of self-representation, but Judge Garland E. Burrell Jr. denied his motion to defend himself by depicting it as a stalling tactic, which forced Kaczynski to be represented by lawyers whose only option to keep him alive was to mount the insanity defense provided by Johnson. Kaczynski thereby found himself confronted with a classic catch-22 situation in which the only way to espouse his views in court was to be labeled insane, and the only way to remain sane was to avoid a trial. Unable to fight the sudden fact of his insane criminality, Kaczynski accepted life in prison in order to avoid the indignity of casting his words and his bombs as the mere ravings and weapons of a lunatic. Unlike August Spies at the Haymarket trial, Kaczynski was denied the chance to explain his revolutionary program, to "call the hangman," and to be martyred for his cause. Thus, the court's refusal to let Kaczynski mount a political defense and its insistence that he was simultaneously insane and competent to both stand trial and represent himself showed that labeling him as insane was a "political diagnosis,"[143] a diagnosis that I suggest derived from the moment his first IED exploded. By using a scientific discipline that can find requisite symptoms to diagnose almost anyone with mental problems, the court used the technoindustrial system to bury Kaczynski's antitechnology terrorism and preempted mass media coverage of the ideas behind his bombs.[144] Kaczynski's attempts to "hit where it hurts" thus empowered the state to reinscribe its own subjugating power.[145] His bombs spoke louder than his words, indicating how the state could muffle Kaczynski's attempt to preserve humanity by destroying it.

In addition to dictating the course of his legal comeuppance, Kaczynski's IEDs triggered a series of systemic repercussions that "entangled" a motley assortment of people in his agenda, "entrapping" them in the FBI's Unabomber investigation, to use archaeologist Ian Hodder's materialist terminology.[146] When unabombs exploded, they motivated specific actions on the part of the technoindustrial system to recover from its brief destabilization. Over seventeen years of fruitless searching, the FBI investigation became broader and broader, sweeping a variety of possible suspects and targets into its purview. The FBI entangled the entire field of STS, going so far as to infiltrate the 1994 Society for Social Studies of Science conference and enlisting STS scholars to help figure out the Unabomber's identity.[147] The list of people who could and might build such bombs was long. Novelist William T. Vollmann became a suspect, owing to loose similarities between his fiction and what little the FBI knew of the terrorist FC, similarities that in the past would have made a hardened conspiracy theorist blush but that now are the banal entanglements potentially faced by everyone living in a hypersurveilled society.[148] Kaczynski's IEDs entrapped certain environmental groups, such as Earth First!, the Earth Liberation Front, green anarchists, and anarchoprimitivists. The state and the mass media used these groups' tenuous association with Kaczynski as a means to justify stigmatizing them as ecoterrorist organizations.[149] These are just a few of the entrapments compelled by his IEDs that did not force a "fight to the finish" with the technoindustrial system.

As Kaczynski's IEDs entangled various people, they also swept technologies into the purview of his antitechnology program, even though, unlike the Luddites, he did not encourage people to smash machines. Yet by threatening the human supporters of machines and systems, unabombs also threatened technologies with destruction.[150] Although he did not aim to blow up technological objects as his primary targets, nonetheless he blew them up. By exploding people and machines, unabombs framed both as equivalent threats to humanity. The various human and nonhuman objects that exploded in unabomb blast radii confused the distinction between the people and the technologies of the technoindustrial system and warranted understanding them as somewhat interchangeable with respect to both the causes and the effects of rhetorical violence. Kaczynski used a bulldozer metaphor to describe how to "hit where it hurts," which demonstrated how technological targets and threats and human targets and threats started to become interchangeable within the context of Kaczynski's terrorism. In short, to destroy a bulldozer (the technoindustrial system), a revolutionary must target the vulnerable engine for attack with the proper tools, and not target the invulnerable blade with a blunt object.[151] Both the bulldozer engine (people) and the bulldozer blade (technologies) threaten humanity because they are operational components of the entire bulldozer (the technoindustrial system). Kaczynski's metaphor clarified that the "vital organs of the system" are not machines. They are people. Machines don't feel pain or coercion, but people do. And the most vulnerable population is the most random—the open target rather than the secure. Instead of sledgehammering computers or blowing them up with IEDs, Kaczynski blew up computer-store workers and computer scientists. Yet at the same time, he also blew up or tried to blow up technological artifacts. When his IEDs exploded, they entangled technologies and superseded his words, which empowered the mass media to charge Kaczynski with being an anachronistically foolish Luddite.[152]

Kaczynski's targeting of the airline industry, its people, and its planes is another example of the interchangeability of people and technologies that the presence of his IEDs entailed. He bombed American Airlines flight 444 in 1979. Twelve passengers were treated for smoke inhalation, and the pilot had to make an emergency landing. Then in 1995, Kaczynski issued a new threat via a letter to the *San Francisco Chronicle*: "WARNING. The terrorist group FC, called unabomber by the FBI, is planning to blow up an airliner out of Los Angeles International Airport some time during the next six days."[153] But with the interchangeability of people and technologies as both threats and targets, the bomb and bomb threat failed to clarify whether the Unabomber's intended target was an airplane, the airliner, its passengers, or the entire industry. The bomb and bomb threat failed to clarify what the aviation industry's invulnerable and nonvital blade and its vital and vulnerable engine were since airliners combine human ingenuity, aviation science, the planes themselves, and the system that sustains them. Hitting an airliner where it hurts—a plane in midflight—established interchangeability between people, machines, and systems via Kaczynski's apolitical terrorism. The attempt to goad a fight to the finish with the airline industry was a lost transmission, disrupted by the ambiguous rhetorical violence of his altitude-sensitive IEDs.

The interchangeability of humans and technologies as targets and threats revealed by Kaczynski's IEDs also demonstrated how rhetorical violence threatened everyone as a universal societal condition of living in a technoindustrial system. Kaczynski's IEDs asked a number of difficult questions about society. As citizens of the technoindustrial system made complicit simply by inhabiting a particular territory and using everyday technologies, are we all targets?[154] Or if we agree with the Unabomber that technology is out of control, are we "all becoming bombs?"[155] What does it mean to be an accomplice to technoindustrial society? Or to live where one's indebtedness to capitalism and its technology make almost everyone potentially complicit in its subversion? The answers to these questions remain elusive. Kaczynski's bombs asked whether people prefer being dominated by technology or being threatened by terrorism. In terms of Technē's Paradox, both technology writ large and the technologies of terrorism threaten destruction and promise preservation. The explosion of a unabomb therefore resembles a question mark more than an exclamation point, and to take Kaczynski and his IEDs seriously entails choosing between unappealing options. The ambiguous purpose of Kaczynski's IEDs, though, lets people know they are unsafe. While human bodies suffer pain and injury, more critical for terrorism is to attack "the enemy's environment" in order to create an "unlivable milieu," in the words of Peter Sloterdijk.[156] Unabombs spread the threat of terrorism across the entire territory that is bounded by the reach of technology. When everyone was a technoindustrial threat and target, then a unabomb was omnipresent in its potential to explode. In this sense, unabombs punctured the American public, who bore witness to each successive event.[157] By proclaiming all people's guilt in their acquiescence to technology's domination of humanity, Kaczynski's IEDs did not prepare the way for revolution. The terrifying accountability shared by almost all people in the technoindustrial system—announced by his IEDs—motivated revulsion toward their creator rather than solidarity. The public found no reason to swap the "unlivable milieu" of technoindustrial society for the "unlivable milieu" of terrorism. The popular revulsion toward Kaczynski's IEDs, an invention of rhetorical violence, showed that the world's population was incapable of antitechnological revolutionary activity.

Unresolved questions about the instability of his talking bombs empowered critics to carve Kaczynski up into two identities, one they could vilify and one with which they could identify. Without the presence of his IEDs, Kaczynski's identity could not, with ease, slip back and forth between being known as a rational mathematician who opposes technology, the mad terrorist Unabomber, and the man who is both. Computer scientist Bill Joy's reaction to the bombings typifies this bifurcation of the culprit into the competing identities of the Unabomber and Theodore J. Kaczynski. "Like many of my colleagues, I felt that I could easily have been the Unabomber's next target," Joy wrote, while also announcing his allegiance with Kaczynski's misgivings about technology.[158] Specifically, Joy agreed with Kaczynski, but disagreed with FC. Other critics noted that Kaczynski's anxiety about technology lent *Industrial Society and Its Future* a nearly ubiquitous popular appeal and an authenticity that can be traced to the classically rebellious American writings of Henry David Thoreau.[159]

"There's a little bit of the Unabomber in most of us," wrote journalist Robert Wright.[160] "Why did someone so like me commit murder?" mused Keith Benson.[161] Daniel Kevles titled a *New Yorker* editorial "E Pluribus Unabomber: There's a Little of Him in Us All." As Baudrillard surmised, the endemic violence of contemporary culture has fomented "that (unwittingly) terroristic imagination which dwells in all of us."[162] Imagining the downfall of a global power structure is easy enough for a society steeped in multiple genres of post-apocalyptic fantasy.

The presence of the terrorist's weapon, however, locates complete societal breakdown within one's living room, office, bus, or marketplace. Kaczynski's logic was "insightful," but "divorced from ethics," according to journalist Kevin Kelly.[163] Although Benson was able to identify in part with Kaczynski, his colleagues saw only a bomb threat. "I was stunned when I was confronted by the acting chair of my department and was profanely accused of 'recklessly endangering the lives of the faculty and staff in my department [who] perceived' they were in danger and, therefore, experienced great discomfort," he wrote of being entangled with the FBI's Unabomber investigation.[164] So as much as people were able to identify at least in part with Kaczynski, they, for the most part, rejected the Unabomber's violence. Thus, the presence of unabombs in a society engulfed by problematic technologies caused a strange fracturing of Kaczynski's identity, whereby everyone is provided with the grounds both to be able to like him and to despise him. When the Unabomber's IEDs punctuated Kaczynski's antitechnology writings, the inseparability of rhetoric and violence was thus less apparent to those who encountered his writings and his bombs as discrete objects.

So for all of Kaczynski's efforts to direct the meaning of Freedom Club's terrorism, the IEDs pushed back, not empowering a withering "fight to the finish," but instead empowering a debilitating psychological diagnosis, the entrapment of random civilians in his investigation, confusion about whether people or technologies are the most vulnerable and vital elements of the technoindustrial system, and the bifurcation of his identity into that of a reasonable skeptic or a despicable terrorist. Thus, his IEDs manifested rhetorical violence, or the unforeseen ways that being in the presence of a weapon entails responses to violence within a system. And Kaczynski's IEDs teach a lesson about how difficult it can be to negotiate Technē's Paradox from within the context of terrorism. Terrorists' weapons materialize Technē's Paradox by bringing world-ending violence into the places where people live, and by sustaining state violence in the name of world-preserving counterterrorism. The presence of a terrorist's weapon is as obligatory as the weapon's capacity to legitimate the suppression of terrorism. For Kaczynski, the IEDs, the objects themselves, seemed to show as much contempt for their creator as Kaczynski had for technology.

3. Conclusion

Regardless of his ability to tap into societal misgivings about technology, Kaczynski, the Unabomber, is an unapologetic terrorist whom most people despise. Hence, I conclude by pointing out how two of the more odious characteristics of his antitechnology agenda are important for the longitudinal history of weapons rhetoric. First, the choice either to submit to technological domination or to revolt against it was not the only irreparable locus that Kaczynski's hard rhetoric made his audience ponder. The dilemma of whether to commit violence also confronts anyone who chooses to side with Kaczynski, to act with neo-Malthusian intent, to take up arms in the vein of Spies's and Fries's advocacy, or to produce the means of violence in the vein of Szilard. The choice to spill blood is its own irrevocable, permanent decision, and sometimes, so too is the choice to promote violence. When he left his first bomb in a parking lot outside the University of Illinois at Chicago's Science and Engineering building, Kaczynski thus faced his own irreparable locus. The violence committed by the Unabomber discomfits those who share Kaczynski's industrial antipathy. Few are the technology haters willing to become terrorists. Many are the technology critics willing to diminish nontechnological problems by insisting on the centrality of technology to human life, to diminish the beneficial aspects of technology, and to amplify the destructive effects of technology. Kaczynski's hard rhetoric, I suggest, incorporates commonplace rhetorical tactics used by rhetors who resist and who dissent against weapons. Kaczynski presented humanity with the irreparable locus of violence, and he offered himself as its terroristic provocateur. And that is exactly what distinguishes his conception of technology from so many philosophers of technology. Kaczynski's rhetorical violence marks him as remarkable and highlights the unlikelihood that antiweapons rhetors will take the step of preparing for battle.

Second, Kaczynski's simultaneous disregard for human life and high regard for the freedom and dignity of humanity manifested as an extreme version of neo-Malthusianism. Kaczynski's neo-Malthusianism, however, exemplifies how governments can lose or cede their thanatopolitical power to authorize the death of particular populations to nonstate actors and terrorists. Kaczynski was neo-Malthusian in the sense that his terrorism aimed at the total annihilation of the technoindustrial system, which would entail an enormous check to global population levels as an unavoidable result of civilization's collapse.[165] Kaczynski often argued that violence is not bad but necessary. When dealing with the technoindustrial system, "non-violence is suicide," he wrote.[166] Yet if his anti-technology program ever succeeds, violence will end up compelling mass suicide as a matter of course. His revolutionary advocacy thereby promoted the lives of a very small minority of survivors above the vast majority, who would neither survive the big technological collapse nor the resulting primitive life. But Kaczynski recommended purging the earth of most humans because of the presence of too many machines and technological systems, not because of the presence of too many humans. The survival of freedom and dignity thus was promoted above the survival of biological life in Kaczynski's version of Malthusian thinking, and technology and weapons would bring destruction in both their presence and absence. In the end, even detestable neo-Malthusian projects such as eugenics and forced sterilization seem more palatable than Kaczynski's global death wish that reminds all people that they are potential targets of technoindustrial, governmental, and nonstate violence.

When Kaczynski's IEDs spoke in his stead, they corroborated the public's negative assessment of his campaign. Like the mustard gas of World War I proved recalcitrant to Fries and West's attempts to proclaim the humanity of chemical warfare, Kaczynski's IEDs proved recalcitrant to his revolutionary message. His eloquent objects communicated the inhumaneness of terrorism rather than the bombs' capacity to catalyze a humane postapocalyptic society; they communicated the indignity of randomly being murdered rather than the dignity of living in what he called "WILD nature"; they communicated the oppression entailed by living under the threat of terroristic violence, rather than freedom from technology.

Kaczynski's case subverts and reaffirms competing reasons to affirm or reject violent revolution, terrorism, and the presence of weapons, all of which makes summing up Kaczynski a somewhat difficult task. His rhetorical self-portrait is blurred.[167] He is not easily categorized because he forces people both to agree with and to disagree with his hard rhetoric, and because he forces people to consider irreparable decisions. "His rhetoric notwithstanding, he was not a technocritic, nor was his primary motivation technosocial criticism in the interest of progressive social change. And strictly speaking, he wasn't an anarchist. He was a mathematician," according to Sal Restivo, whom the FBI enlisted to help locate Kaczynski.[168] But he wasn't exactly a mathematician. Or a philosopher of technology. Or a Luddite. And he wasn't exactly a neo-Malthusian. Even to call Kaczynski a terrorist does not quite fit his rhetorical self-portrait. The old cliché that one person's terrorist is another person's freedom fighter tramples upon the nuance and complexity of armed conflict, yet it seems like a fitting piece of the legal defense that he was never allowed to mount. Although no one authorized him to do so, and some might despise him for it, he murdered people to liberate people from technology. And his bombs say that he was insane and, therefore, easily dismissed. To take Technē's Paradox seriously, though, one must also take into account Kaczynski, his bombs, and his unwavering assault on the idea that technology is preservative.

Conclusion | In the Presence of Weapons and Rhetoric

The point is not that atomic weapons constitute a new argument. There have always been good arguments.
—J. Robert Oppenheimer (1945)

The good craftsmen seemed to me to go wrong as much as the poets: because they practiced their *technai* well, both thought themselves wise in other, most important things, and this error of theirs obscured the wisdom they had.
—Plato (late fourth century BCE)

This book has focused on the interactivity of weapons and rhetoric at the nexus of Technē's Paradox. This commonplace paradox simultaneously holds that technology will destroy humanity, if not cause its total extermination, and that technology will also preserve humanity from destruction.

Technē's Paradox, as I asserted in the introduction, is a commonplace way of thinking and talking about all dangerous technologies as well as a guiding logic for thinking about much less dangerous technologies, and even the ordinary technologies we unthinkingly encounter every day. The dual uses of nitrogen fertilizers (introduction), Thomas Malthus's portrayal of agriculture as preservative and destructive (chapter 1), and Ted Kaczynski's denunciation of all technologies as destructive despite their apparent benefits (chapter 5) gesture toward the sweeping relevance and influence of Technē's Paradox. Controversial technologies, especially, seem to keep promising both preservation and destruction. And, whether some predictions end up being incorrect, remain inconclusive, or seem overexaggerated, Technē's Paradox maintains its durable capacity to motivate belief and action. Although the apocalyptic prophecy proved false, rhetors negotiated Technē's Paradox, for instance, when they debated whether the European Council for Nuclear Research's Large Hadron Collider would, in the quest to create the Higgs boson (the "God Particle"), blow up the earth by initiating a second big bang. The Higgs boson was observed, and hope remains that such research might aid humanity by divulging some mysteries of the universe.[1] While competing evidence suggests that artificial intelligence might save humanity by outperforming the human brain and might command its autonomous robots to run amok in human blood, both possibilities are, at the moment, speculations.[2] Rampant hyperbole about algorithmic computer software's dual capacity to preserve and destroy individual lives can be observed throughout society, from smart phones' alluring double power to turn people into unthinking zombies and to assist with every possible preservative activity, to the computerized gambling machines that promise a life-preserving jackpot but most often deliver financial ruin.[3] In both cases, empirical evidence grants plausibility to hyperbolic encomiums and invectives directed at digital culture. Technē's Paradox remains vital to technology rhetorics, and rhetors must continue to negotiate it.

The life-and-death effects of killing machines are more palpable and urgent than those of particle colliders, sci-fi robots, and smart phones, though. Hence, throughout the book I showed that the binary poles of Technē's Paradox affirm the inseparability of weapons and the language used to advocate and resist them, and that weapons provide a particular and important insight into how humanity imagines its collective technological fate. The importance of Technē's Paradox, weapons, and rhetoric derives from their interaction whenever people hover in the balance between life and death. Weapons rhetoric functions at the tipping point between preservation and destruction, while the negotiation of Technē's Paradox points humanity toward communication or killing, diplomacy or war, political reforms or revolutionary terrorism, and the ultimate, exaggerated results of utopia or extermination.

The preceding chapters followed the longitudinal history of weapons rhetoric in order to examine both what being in the presence of weapons entails, and what being in the presence of words about weapons entails. One clear entailment is that weapons rhetors tend to negotiate the Paradox in one way or another—by engaging it directly, by focusing attention on one pole or the other, or by attempting to subvert it. The analysis in chapter 1 of Thomas R. Malthus's *Essay on the Principle of Population* and its portrayal of the population bomb served as a pre-text for why global destruction and preservation came to inhere as primary commonplaces in weapons rhetoric. Malthus negotiated Technē's Paradox by providing some stock tactics of weapons rhetoric—sublime statistical manipulation, amplified universal generalizations, and machine *antistasis*—

which pushed people to think about how the promise of utopia pushes humanity to the brink of extinction. Chapter 2 showed that August Spies negotiated Technē's Paradox in his "Address to the Court" by exploiting the rhetorical instability afforded by the presence of dynamite bombs with strategic polysemy and turnaround arguments that affirmed dynamite's simultaneously destructive and preservative powers. Chapter 3 showed how Maj. Gen. Amos A. Fries and Clarence J. West negotiated Technē's Paradox by attempting to establish chemical weapons as categorically preservative, which negated their destructive capacity. They advocated the preservative power of chemical weapons with statistical proofs of their humaneness and totalizing, hyperbolic amplification of their strategic uses. Chapter 4 demonstrated how Manhattan Project physicist Leo Szilard negotiated Technē's Paradox by living it. His rhetorical being combined scientific and political *ethoi*, embodied a collective scientific voice, and crafted an atomic realpolitik attitude to lead a paradoxical life. Chapter 5's examination of Unabomber Ted Kaczynski's hard rhetoric showed how he diminished non-technological problems, diminished technological advantages, and amplified technological destruction, oppression, and domination in order to negotiate Technē's Paradox by categorically proclaiming all technology destructive. All together, the eleven rhetorical tactics of weapons rhetoric I have examined are the use of sublime statistics, universal generalizations, *antistasis*, strategic polysemy, turnaround arguments, statistical manipulations, amplification, diminishment, the manipulation of *ethoi*, collective rhetorical embodiment, and realpolitik machinations.

Of course, many more rhetorical tactics and strategies exist within the material contingencies of weapons rhetoric. Nonetheless, when weapons rhetors confront Technē's Paradox again and again, certain commonplace tactics and overriding rhetorical strategies emerge and are reiterated such that the longitudinal history of weapons rhetoric suggests a dominant network of weapons discourse. The rhetorical tactics and strategies associated with the Bomb, modern-day terrorism, and drone warfare have much older origins, and they help to invent the material, political, cultural, and ideological presence of weapons.

This dominant network of weapons discourse mobilizes some rhetorical elements, such as amplification and hyperbole, which are constant across the case studies and the development of modern warfare. The amplification used to describe the destructive power of different weapons, for instance, is more or less hyperbolic depending on circumstances. Whereas a dynamite bombing killed one person at Haymarket, one mustard-gas shelling decimated an entire battalion at Ypres, and the nuclear bombings of Hiroshima and Nagasaki killed hundreds of thousands of city dwellers, all three weapons were described as capable of destroying the entirety of human civilization. Hyperbolically, Kaczynski urged his readers to think of his bombs as catalysts for an antitechnology revolution that would destroy all systems and machines in order to preserve humanity. The destructive scope of the next Malthusian population bomb seems limited only by humanity's technical capacity to preserve itself by engineering and limiting catastrophe. Hyperbole and amplification appear in an ever-changing array of different forms.[4] Hyperbole and amplification have thus been used for a variety of political, military, and social circumstances from instilling revolutionary fortitude in workers to justifying the "military-industrial complex."[5] Thus, over time, the example of hyperbole shows that the ways of describing, advocating, and resisting weapons have coalesced into an observable template of rhetorical choices that is added to and amended according to varying contexts. The relative constancy of such weapons rhetoric derives, I have argued, in part from the ubiquitous presence of the technological conundrum I call Technē's Paradox.

The dominant network of weapons discourse sometimes dictates politics, thereby hijacking communication in order to subordinate communication to a weapon. If people believe weapons advocates, then political expediency must be oriented around the likelihood of whether people will resort to killing each other with particular weapons. Weapons of mass destruction have justified a wide range of US foreign policies, such as threatening North Korea, waging war against Iraq, and maintaining the Cold War. The Bomb's geopolitical domination of Cold War political discourse compelled Hannah Arendt, for one, to write, "the great political issue of freedom versus total domination is overshadowed by the fear of extinction."[6] Because the Bomb materialized and amplified a ubiquitous technological anxiety, the weapon's influence was traceable within Cold War discourse as a manifestation of Technē's Paradox. In a type of technological circular reasoning, the Bomb became both the means of conducting foreign policy and the indispensable proof of the reasons for enacting those policies.

Technē's Paradox teaches other lessons that help people learn to live with weapons without fretting about them all the time, whether with naïveté or vigilance. In its ubiquitous insistence, Technē's Paradox presents a false dilemma that prescribes only two results—utter apocalyptic destruction and purified utopian visions—both of which remain fantastical. False dilemmas, however, reveal that more-moderate paths of action exist to deal with our collective technological messes both conceptually and practically, whether through risk analysis or by conceiving of ways to reconfigure the relationship of technology to the economy. If it does not cause relief, then the history of weapons rhetoric teaches humanity to question appeals to progress, appeals to technological ingenuity, and appeals to technological solutions for technological problems, all of which can mask or "sterilize" weapons-related violence.[7]

Paradoxical weapons rhetoric helps the public to imagine weapons so that incongruous practical and ideal political and military goals seem capable of equal, contemporaneous achievement. Paradoxical weapons rhetoric also helps individuals and populations understand how they live and think while designing, manufacturing, advocating, deploying, ignoring, fearing, reviling, and resisting weapons. Paradoxical weapons rhetoric contributes to the invention of a shared global technological condition visible across different times, places, events, and discourses.[8] This technological condition and paradoxical weapons rhetoric have a continuing history. They neither originated nor ceased when people seemed to become terrified of looming global nuclear holocaust. The Bomb just integrated older patterns of weapons rhetoric into a new assemblage. And more rhetorical weapons assemblages continue to manifest, as evidenced by contemporary drone-warfare discourse, in which drone advocates and drone detractors reiterate and negotiate the paradoxical rhetorical commonplaces of weapons rhetoric.[9]

Yet as much as the longitudinal history of paradoxical weapons rhetoric indicates a dominant discourse and certain rhetorical commonplaces, vibrancy and change also characterize it. The presence of new weapons alters weapons rhetoric when rhetors calibrate their language to new inventions, changing political contexts, and the sizes of populations that must be preserved or annihilated with war and terrorism. If the Bomb did not shatter rhetoric at first blast, as I have suggested in contradiction to other assessments of nuclear-weapons rhetoric, then, still, it did perturb it.[10] Despite its constancy over time, weapons rhetoric demonstrates novelty as much as normalcy.

The novelty and change evident in weapons rhetoric offers some hope that rhetoric might be the means to begin refusing, chipping away at, undermining, and attacking the armed technological society and its dominant network of discourse. If humanity must resist dangerous technologies and work to solve the damage wrought by the presence of weapons, then it can use rhetoric to invent the means to do so. Janet Atwill asserted that the concept of technē is imbued with this transformative power. "Technē challenges those forces and limits [compulsion, necessity, and fate] with its power to discover and invent new paths," and technē "makes intervention and invention possible," she wrote.[11] If the technē of rhetoric entails the art of discovering the available means of persuasion,[12] then in a society that is integrated with weaponry, that art must persuade by negotiating both the philosophical and ideological concepts of society that define it as technological, as well as the preponderance of weapons that make their own arguments. Empowered by technē, critics can help elucidate society's responses to a world that is only partly subjugated to weaponry. People have the freedom to resist weapons and communicate their dissent to others. But the *technai* of rhetoric and weapons helped to put humanity into its current technological predicaments in the first place. There have always been practical reasons to make more fertilizer and more explosives. As entrenched societal *technai*, rhetoric and weapons might inhibit the discovery of nonviolent solutions rather than provide them. Plato's Socrates said, "The good craftsmen seemed to me to go wrong as much as the poets: because they practiced their *technai* well, both thought themselves wise in other, most important things, and this error of theirs obscured the wisdom they had."[13] Dissenting weapons rhetoric might prove to be a limiting force rather than a liberating one, obscuring its own wisdom by obscuring other rhetorical or nonrhetorical means of resistance.

In what I have called this book's middle-way sections, I have shown how the presence of weapons entails that people cannot always use rhetoric to control their inventions, and that weapons possess a certain persuasive power unto themselves. I have examined five material contingencies entailed by being in the presence of weapons—the motivational force of large populations armed for war when the population bomb looms, the capacity of weapons to prove legal accountability, the recalcitrant capacity of weapons to goad people into describing these weapons in different ways, the verification of speculation based on the presence of a weapon, and the capacity of weapons to speak for people.

Weapons are eloquent objects. For example, Oppenheimer, speaking to Los Alamos scientists in late 1945, indicated that the Bomb possessed a persuasive capacity beyond its technical capacity to destroy, a capacity that complemented its functional military purposes. "The point is not that atomic weapons constitute a new argument. There have always been good arguments," he said.[14] Ernst Jünger said much the same of machine guns.[15] The atomic bomb for Oppenheimer and the machine gun for Jünger were themselves the argument, self-contained and in no need of embellishment and human mediation to help them make their case. So, too, for Malthusian population bombs, dynamite bombs, mustard-gas shells, and IEDs. Every weapon, Oppenheimer suggested, has always made good arguments. And bad ones. The threat of population bombs has justified barbarous mistreatment of populations deemed redundant by states, societies, and religions, while also threating such institutions with eventual retribution. The presence of dynamite linked Spies to anarchism, doomed his legal defense, and blemished labor activism. Chemical weapons ensconced their own antitechnological arguments as social knowledge only at the cost of severe mental and physical trauma. Nuclear weapons terrorize the mind whenever geopolitics turns the world's attention to their unending presence. Ted Kaczynski's mail bombing of David Gelernter was an anti-computer argument with or without the presence of *Industrial Society and Its Future*. Thus, each case showed how weapons displaced the agency of human weapons advocates, and often motivated more revulsion than approval of the weapons. The presence of mustard gas and unabombs, for instance, inspired little pro-chemical-warfare sentiment and little antitechnological fortitude. The arguments made by any weapon's material presence spin out of the control of the weapon's designers, manufacturers, strategists, users, advocates, and detractors, just as the arguments made by people's words spin out of their control.

The rhetorical presence of a weapon raises a difficult challenge for rhetors who want to overturn and refute the opposition's arguments when the opposition's arguments are warranted by violence and death. The presence of weapons presents a particular conundrum that makes it difficult to disentangle how motivation, persuasion, and causation materialize when both weapons and words are complicit in violence. Fries and West, for instance, faced this conundrum when they attempted to sell the banality and effectiveness of gas defensive training to soldiers who were living the abnormality of World War I trench warfare and the incapacity of high command to protect soldiers. That the National Rifle Association can keep its model of gun regulation in place in the wake of so much American gun violence is a monumental and dark achievement, an achievement that demonstrates the vast power of rhetoric mixed with money and weapons. If greater proximity to weapons grants them greater rhetorical presence, then a related challenge for rhetors who debate weapons policy, whether for the military, the police, citizens, or the resistance, is to give presence to weapons when they are absent.[16] The absence and presence of weapons, as well as the presence of absent weapons in language, demarcates and limits the capacities of people to be violent and to be safe.

Although weapons advocates can downplay a weapon's destructive power, the inventiveness of rhetoric, technologies, and concepts cannot hide the killing force of weapons, which reveals that people have justified reasons to question, wonder about, and fear weapons that could end up annihilating their worlds, whether that world is the site of a mass shooting, a city in Japan, or, in our imagination, the entire world. Technē's Paradox, as it confronts people with all of its commonplaces, quirks, contradictions, and complexities reveals that even powerful individuals and institutions struggle to control the simultaneously preservative and destructive presence of weapons. Much like the fraught tension at the limit between violence and rhetoric, Technē's Paradox reveals how language might fail to exercise control over a dangerous technology just at the moment when violence looms, or how language might help to avert disaster. At this palpable limit between war and peace, Technē's Paradox demonstrates whether the action a rhetor must take is one of communal preservation, collective destruction, or both simultaneously.

When a weapon's presence and the threat of violence loom, the urgency of rhetoric is at its utmost. Yet in spite of the urgency projected by the *Bulletin of the Atomic Scientists*' "doomsday clock," which year after year counts down the final minutes until globalized technological annihilation, we wait around for the ultimate answers to the questions provoked by the presence of ultimate weapons and the most dangerous technologies.[17] Is all life poised for total annihilation or total preservation? In order to communicate maximum urgency, the doomsday clock's minute hand is always poised at just minutes to midnight. But, as the Italian thinker Elias Canetti remarked, "the more distant the goal, the better the prospect of its permanence."[18] If we question whether humanity's fate is either technological annihilation or technological preservation, then we'll wait until the end of time to know the answer.

Notes

Introduction

The epigraphs for this chapter are drawn from Jünger, "Technology as the Mobilization of the World," 289, and "Wo aber Gefahr ist, wächst / Das Rettende auch" (Hölderlin, *Sämtliche Werke*, 1:350), also quoted in Heidegger, *Question Concerning Technology*, 28 and 34.

1. Rather than "preservation," technology scholars and critics often use the term "salvation" to refer to the beneficial effects of technology. Deifying technology to explain the secularization of modern society has become an oft-named symptom of modernity—and hence a common *topos* for technology critics who examine the sudden proliferation of technology and industrialization and the decline of religious influence in the 1700s and 1800s (see Ellul, "Technological Order," 96; Mumford, *Myth of the Machine*, 4, 45, 53–56, and 365; and Noble, *Religion of Technology*, 208). Using preservation instead of salvation indicates that Technē's Paradox confronts people at all stages of history, regardless of religious climate.

2. "Background technology" is philosopher Don Ihde's term for technologies that people tend to not notice as long as they operate properly (*Technology and the Lifeworld*, 108–12).

3. Mumford, *Myth of the Machine*, 55. See also Holley, *Ideas and Weapons*, 175.

4. Aristotle, *On Rhetoric*, 1354a–1355b. See Janet Atwill's *Rhetoric Reclaimed* for an extensive etymology of the term technē and a history of its usage.

5. Nussbaum, *Fragility of Goodness*, 95.

6. Nussbaum, *Fragility of Goodness*, 95.

7. Nussbaum, *Fragility of Goodness*, 95.

8. Writing about technology and rhetoric together creates difficulties for defining the two terms, while the concept of technē brings rhetoric and technology into alignment. Defining both rhetoric and technology as techniques, as Ellul did with his concept of *la technique* in *The Technological Society*, also aligns the two terms via their overlapping connotations of organized, inventive productivity. In this first sense, both technologies and the uses of language are techniques. For the purposes of this book, readers should keep in mind the overlapping characteristics of technology and rhetoric as inventive techniques, but I also want readers to consider the profound difference that sometimes exists between a technology and words about it. Sometimes a technology and words about that technology appear together as a unified technique, together in a particular context, or together as related but discreet techniques in an assemblage. At other times, a technology and words about that technology do not appear together. A technology and words about that technology can be separated by time, geographical distance, media, tropes, semiotics, dissociation, and fidelity to reality. To complicate matters even further, all of these instances of differentiation could also be framed as part of an all-encompassing assemblage or technique. Yet, in this book, I want to emphasize that there is sometimes a distinct difference between being in the presence of a weapon and being in the presence of words about a weapon. For instance, there is a distinct difference between being trapped in a room with a ticking time bomb and watching a movie about people being trapped in a room with a ticking time bomb. The actual time bomb and the movie time bomb will function rhetorically in very different ways to motivate belief and

action because they rely on different kinds of presence. In this second sense, technology can refer to machines, instruments, or physical apparatuses, and rhetoric can refer to words, language, symbols, representation, argumentation, and communication. I thus rely on both the overlapping definition of technology and rhetoric as techniques of motivating belief and action, and the dissociation of their definitions that specific contexts and exigencies sometimes necessitate.

9. See McKeon's "Uses of Rhetoric in a Technological Age" (2 and 12–13); Kennedy's commentary on Aristotle (Aristotle, *On Rhetoric*, 12–13); and Kennedy's *Classical Rhetoric* (21). In *Metaphysica*, Aristotle linked *technē* and *logos* to help define human existence. He wrote that, in addition to experience, "the human race lives also by art [*technē*] and reasoning [logos]" (689 [980b]).

10. Elsewhere, in "'The Human Barnyard' and Kenneth Burke's Philosophy of Technology," I have written about how, throughout Burke's writings, he imbued technology with rhetorical capacities.

11. Latour, *Pandora's Hope*, 176–80.

12. Hill, "Memes, Munitions, and Collective Copia."

13. Keränen, "Concocting Viral Apocalypse"; Mitchell, *Strategic Deception*; Schiappa, "Rhetoric of Nukespeak."

14. Virilio, *Information Bomb*, 63; Virilio, *City of Panic*, 32.

15. In "Why Satire, with a Plan for Writing One," Kenneth Burke noted that people tend to think according to the Aristotelian concept of "entelechy," which Burke defined as "tracking down the implications of a position, going to the end of the line" (29).

16. Oklahoma Today, *9:02 a.m. April 19, 1995*, 15.

17. Michel and Herbeck, *American Terrorist*, 215.

18. F. Haber, "Synthesis of Ammonia from Its Elements," 14.

19. Charles, *Master Mind*, xi. Worldwide nitrogen production from the Haber process increased from 6,798 tons in 1913 to 297,000 tons in 1920 (Haynes, *American Chemical Industry*, vol. 2, 362).

20. Johnson and MacLeod, "War the Victors Lost."

21. According to one 2005 study, approximately two billion people would starve without nitrogen fertilizers (Charles, *Master Mind*, xiv).

22. Plato, *Protagoras*, 319.

23. Plato, *Protagoras*, 320.

24. Ellul, "Technological Order," 97–98.

25. Ellul, "Technological Order," 97.

26. Collins and Pinch, *Golem at Large*, 1.

27. R. B. Fuller, *Utopia or Oblivion*, 291–92.

28. Beck, *World Risk Society*.

29. Beck, *Ecological Enlightenment*, 22–23.

30. Beck, *Ecological Enlightenment*, 2.

31. Erasmus, *De Copia*, 303.

32. Keller, *Making Sense of Life*, 120, 118, 168, and 195.

33. Keller, *Making Sense of Life*, 167 and 152.

34. Keller, *Making Sense of Life*, 170.

35. Foucault, *Birth of the Clinic*, xvi–xvii, translation altered.

36. J. Campbell, "Between the Fragment and the Icon," 347 and 346.

37. See McGee's "'Ideograph'" and Leff's "Things Made by Words."

38. J. Campbell, "Between the Fragment and the Icon," 348.

39. J. Campbell, "Between the Fragment and the Icon," 353 and 364.

40. J. Campbell, "Between the Fragment and the Icon," 365.

41. J. Campbell, "Between the Fragment and the Icon," 367.

42. J. Campbell, "Between the Fragment and the Icon," 350–51.

43. This formulation is in no way meant as a negative critique of two alternate approaches to the inseparability of rhetoric and technology: one based in social constructivism that conceives of technologies as texts to be interpreted (Woolgar, "Turn to Technology," 39), and one that understands rhetoric as a technology (Barthes, *Semiotic Challenge*, 50; Bazerman, "Production of Technology and the Production of Meaning," 383; J. Brown, "Machine That Therefore I Am"; Greene, "Another Materialist Rhetoric," 21; Ong, *Rhetoric, Romance, and Technology*, 4).

44. Graham, *Politics of Pain Medicine*, 12; Harman, "Vicarious Causation," 190; Jasanoff, "Ordering Knowledge, Ordering Society," 14; Bennett, *Vibrant Matter*, viii and 1–19.

45. Latour, *Pandora's Hope*, 190. I do not seek any methodological authorization from any particular materialist approach, but I do mobilize insights from object-oriented ontology, speculative realism, new materialism, and other materialist approaches.

46. S. Fuller, *Philosophy, Rhetoric, and the End of Knowledge*, 36; Graham, *Politics of Pain Medicine*, 8–16.

47. Latour, *Pandora's Hope*, 95. Networks, or collectives of "facts, power, and discourse" propagated by both human and nonhuman "actants," stimulate people to respond (*We Have Never Been Modern*, 6). A Latourian "network of power" has "properties borrowed from the social world in order to socialize and properties borrowed from nonhumans in order to naturalize and expand the social realm" (*Pandora's Hope*, 204).

48. Lynch and Kinsella, "Rhetoric of Technology"; Kling, "Audiences, Narratives, and Human Values," 355.

49. Shaffer, "Science Whose Business Is Bursting," 148.

50. Shaffer, "Science Whose Business Is Bursting," 148; Daston, "Introduction: Speechless," 12–13.

51. Bryant, Srnicek, and Harman, "Towards a Speculative Philosophy," 3.

52. Barad and Daston are not alone when it comes to STS approaches that resemble Campbell's proposed middle-way path between texts and materialism. Theories of human-technology hybridity (Haraway, *Simians, Cyborgs, and Women*, 150), heterogeneity, monstrosity (Law, "Introduction," 10–11 and 16–17), and posthumanity (Hayles, *How We Became Posthuman*)—which are discursively and materially observable in assemblages, networks, and "sociotechnical systems" (Latour, *We Have Never Been Modern*, 3–5)—grant the force of motivation to both human and nonhuman entities, individually and together.

53. Barad, *Meeting the Universe Halfway*, 3.

54. Barad, *Meeting the Universe Halfway*, 136.

55. Daston, "Introduction: Speechless," 10.

56. Daston, "Introduction: Speechless," 12.

57. Daston, "Introduction: Speechless," 20.

58. Bryant and Joy, "Preface," ix.

59. Bryant and Joy, "Preface," iv.

60. Lazzarato, *Signs and Machines*.

61. Simons, "Introduction," 1; Bennett and Joyce, *Material Powers*, 4; Bryan, Srnicek, and Harman, "Towards a Speculative Philosophy," 1; Simonson, "Reinventing Invention, Again," 313; Milbourne and Hallenbeck, "Gender, Material Chronotypes, and the Emergence of the Eighteenth Century Microscope," 402; Rivers and Weber, "Ecological, Pedagogical, Public Rhetoric"; May, "Orator-Machine"; Pickering, *Mangle of Practice*, 21 and 7; Akrich, "De-Scription of Technical Objects."

62. Rickert, *Ambient Rhetoric*, 145 and 160; Gries, "Dingrhetoriks," 301–4.

63. McGee, "Materialist's Conception of Rhetoric," 10 and 28–29; McKeon, "Uses of Rhetoric in a Technological Age," 18.

64. McKeon, "Uses of Rhetoric in a Technological Age," 18; S. Fuller, "Strong Program in the Rhetoric of Science," 96–97.

65. Rhetoricians have tinkered with the traditional concept of rhetorical agency both with and without recourse to STS scholarship. See C. Miller, "What Can Automation Tell As about Agency?"; K. Kennedy, "Textual Machinery," 304; Doyle, *On Beyond Living*, 6 and 26; and Rickert, *Ambient Rhetoric*, xv. On the persuasiveness of objects, see Hallenbeck, "Toward a Posthuman Perspective"; and Graham, "Agency and the Rhetoric of Medicine."

66. Morton, "Sublime Objects," 209. Jane Bennett defined rhetoric as "word sounds for tuning the human body; for rendering it more susceptible to the frequencies of the material agencies inside and around us" ("Powers of the Hoard").

67. Hughes, "Technological Momentum," 104.

68. Bennett, *Vibrant Matter*, 122.

69. See Morton, "Sublime Objects," 213.

70. Brown and Rivers, "Composing the Carpenter's Workshop," 29.

71. Perelman and Olbrechts-Tyteca, *New Rhetoric*, 116, emphasis in original; Locke, *Essay Concerning Human Understanding*, 109.

72. Perelman and Olbrechts-Tyteca, *New Rhetoric*, 117.

73. Perelman and Olbrechts-Tyteca, *New Rhetoric*, 117–18.

74. Virilio, *War and Cinema*, 8.

75. "Solutions to the problem of knowledge are solutions to the problem of social order," posited Steven Shapin and Simon Shaffer (*Leviathan and the Air-Pump*, 332). The authors called this statement their "methodological slogan" (xl).

76. For an examination of Malthus's term "redundant population," see Hill's "Rhetorical Transformation of the Masses."

77. Lanham, *Handlist of Rhetorical Terms*, 16.

78. W. Benjamin, "Theses on the Philosophy of History," 256–57.

Chapter 1

The first epigraph to this chapter is drawn from Malthus's second *Essay*, vol. 2, 185. The second epigraph is quoted in Bell, *First Total War*, 206 and 251. Bonaparte supposedly said this to Austrian statesperson Klemens Wenzel Nepomuk Lothar von Metternich.

1. In the notes to this chapter, I follow the trend in Malthus scholarship by referring to the 1798 edition as the "first *Essay*," and by referring to the 1803 and subsequent editions as the "second *Essay*." The reason for the distinction between the 1798 and latter editions rests on Malthus's almost complete revision of the first edition in 1803, when he excised the majority of the original text, revised what he retained, and added hundreds of pages of additional evidence. Less substantial revisions appeared in 1806, 1807, 1817, and 1826. His brief final statement on the subject, *A Summary View of the Principle of Population*, was a shortened version of an entry on population he wrote for *Encyclopedia Britannica* in 1830. In the main text of this chapter, I refer to all of Malthus's writings about population theory collectively either as the *Essay* or the *Principle of Population*.

2. Malthus, first *Essay*, 72.

3. Stangeland, *Pre-Malthusian Doctrines of Population*, 353–54.

4. My recounting of Malthus's life and activities relies on Patricia James's *Population Malthus*.

5. Digby, "Malthus and Reform of the Poor Law," 104.

6. Digby, "Malthus and Reform of the Poor Law," 104.

7. London seemed to be "the great devourer of lives from the country, a town that absorbed an inordinate part of the country's natural population increase and proceeded to kill it" (Schwarz, *London in the Age of Industrialization*, 237).

8. Bonar, *Malthus and His Work*, 174, 304–5, and 317; Digby, "Malthus and Reform of the English Poor Law," 167; Huzel, *Popularization of Malthus*, 3; Meek, *Marx and Engels on the Population Bomb*, 8; Patriquin, *Agrarian Capitalism and Poor Relief*, 15; K. Smith, *Malthusian Controversy*, 296. For counterarguments that Jeremy Bentham was as influential as Malthus, if not more so, regarding the 1834 Poor Law reformation, see Charlesworth, *Welfare's Forgotten Past*, 14; Brundage, *English Poor Laws*, 32–36, 40, 45–48, and 54; Digby, "Malthus and Reform of the Poor Law," 105.

9. Brundage, *English Poor Laws*, 12 and 79–81.

10. Malthus, second *Essay*, I, 13; Malthus, *Occasional Papers*, 140; Winch, "Introduction," ix–x.

11. Bonar, *Malthus and His Work*, 1.

12. Smith, *Malthusian Controversy*, 48.

13. An extensive bibliography of contemporary responses to Malthus by other famous period writers, like William Godwin, William Cobbett, and William Hazlitt, can be found in D. V. Glass's *Introduction to Malthus* (79–112). For the most persistent criticisms of Malthus, see Avery, *Progress, Poverty and Population*; Beales, "Historical Context of the *Essay* on Population"; Bonar, *Malthus and His Work*, 355–98; Meek, *Marx and Engels on the Population Bomb*; Petersen, "Malthus and the Intellectuals"; and Smith, *Malthusian Controversy*, 47–206.

14. Malthus, second *Essay*, I, 21–22.

15. Malthus disparaged the "absurd paradoxes" of William Godwin, the Marquis de Condorcet, and other utopian supporters of the French Revolution (first *Essay*, 68 and 130). Rather than avoid paradoxes, though, Malthus depended on his own erudite use of them to explicate his population theory. As Bonar noted in his nineteenth-century interpretation of Malthusianism, "The main position of the [*Essay*] was so incontrovertible, that when the critics despaired to convict Malthus of a paradox, they charged him with a truism" (*Malthus and His Work*, 85). For more on Malthus's use of paradox, see Bonar, Fay, and Keynes, "Commemoration of Thomas Robert Malthus," 62; Dupâquier, *Malthus Past and Present*, viii; Hodgson, "Malthus," 3; Hollander, *Economics of Thomas Robert Malthus*, 950; James, *Population Malthus*, 109; Pullen, "Malthus on the Doctrine of Proportions," 430; and Winch, *Malthus*, 6 and 35.

16. Zarefsky, *Rhetorical Perspectives on Argumentation*, 91.

17. Farrell, "Knowledge, Consensus, and Rhetorical Theory," 4.

18. Hasian, "Legal Argumentation in the Godwin-Malthus Debates," 194.

19. Hacking, *Social Construction of What?* 21.

20. Ehrlich, *Population Bomb*.

21. For a representative extension of Malthusianism to a range of paradoxically annihilating and preservative technological disasters and solutions, see Brown, Gardner, and Halweil's *Beyond Malthus*.

22. Canetti, *Crowds and Power*, 67–73.

23. Corbett, *Selected Essays*, 105.

24. Hawhee, *Bodily Arts*, 10.

25. Hawhee, *Moving Bodies*, 2.

26. Scarry, *Body in Pain*, 125 and 116.

27. Fountain, *Rhetoric in the Flesh*, 14.

28. Bennett, *Vibrant Matter*, 112, emphasis in original.

29. Malthus, first *Essay*, 71, paragraph arrangement altered.

30. Ricardo, *Works and Correspondence*, vol. 8, 361, emphasis in original.

31. Foucault, *History of Sexuality*, 140; Foucault, *"Society Must Be Defended,"* 246.

32. Foucault, *Birth of Biopolitics*, 320. See also Foucault, *"Society Must Be Defended,"* 243.

33. Foucault, *Birth of Biopolitics*, 21.

34. Foucault, *History of Sexuality*, 140 and 141; Foucault, *"Society Must Be Defended,"* 242.

35. Stormer, *Signs of Pathology*, 34.

36. Agamben, *Homo Sacer*, 122. Regarding thanatopolitics, "death gains the upper hand in biopolitical reflection precisely at the moment when questions of technology grow in importance" (T. Campbell, *Improper Life*, viii).

37. Foucault, *"Society Must Be Defended,"* 247; Agamben, *Homo Sacer*, 122–23.

38. Malthus, first *Essay*, 89–90; Malthus, second *Essay*, I, 18. According to Hacking, "statistics, in that period, was called a moral science: its aim was information about and control of the moral tenor of the population ("Biopower and the Avalanche of Printed Numbers," 67).

39. Hacking, "Biopower and the Avalanche of Printed Numbers."

40. Malthus, second *Essay*, I, 18. In the second *Essay*, Malthus excised two items, "vicious customs with respect to women" and "luxury," from the first *Essay*'s much shorter list of positive checks (First *Essay*, 103).

41. On the concept of the technological mastery of nature, see Winner, *Autonomous Technology*, 132.

42. See Isaacson, *Malthusian Limit.*

43. Perelman and Olbrechts-Tyteca, *New Rhetoric*, 116, emphasis in original.

44. Foucault, *History of Sexuality*, 137.

45. Quoted in Bell, *First Total War*, 206 and 251.

46. Aberth, *From the Brink*, 63; Cunningham and Grell, *Four Horsemen*, 98; McNeill, *Pursuit of Power*, 185.

47. Bell, *First Total War*, 115, 252–53 and 296–97; Smith, *Wealth of Nations*, 397.

48. Brodie and Brodie, *From Crossbow to H-Bomb*, 30.

49. Quoted in McNeill, *Pursuit of Power*, 192.

50. Bell, *First Total War*, 124 and 174.

51. Georges Danton, quoted in Bell, *First Total War*, 117. On the polysemous character of the term "total war," see Chickering's "Total War" (18–19).

52. Clausewitz, *On War*, 717.

53. Bell, *First Total War*, 149; Brodie and Brodie, *From Crossbow to H-Bomb*, 106–7.

54. Bell, *First Total War*, 161.

55. Bell, *First Total War*, 212–17.

56. Malthus, second *Essay*, 222.

57. Reproduced in Malthus, first *Essay*, 231.

58. Huzel, *Popularization of Malthus.*

59. Ricardo, *Works and Correspondence*, vol. 8, 107–8.

60. Malthus, second *Essay*, II, 123.

61. See Greene, *Malthusian Worlds*, 12.

62. Foucault, *History of Sexuality*, 143.

63. Foucault, *"Society Must Be Defended,"* 257.

64. Agamben, *Homo Sacer*, 3.

65. Foucault, *"Society Must Be Defended,"* 249.

66. Foucault, *"Society Must Be Defended,"* 258–61; Foucault, *History of Sexuality*, 137.

67. Malthus, second *Essay*, II, 134. See also Malthus, first *Essay*, 237.

68. Malthus, second *Essay*, I, 229. More than three hundred thousand French soldiers died in the battle of 1812 (Bell, *First Total War*, 258).

69. Malthus, *Occasional Papers*, 66.

70. Burke, *Rhetoric of Motives*, 106.

71. Like many of Malthus's claims about population, he was not the first to note the apparent geometrical and arithmetical increases of population and subsistence, and the ratios are comparable to one of Francis Bacon's double arguments (Stangeland, *Pre-Malthusian Doctrines of Population*, 354; Bacon, *Advancement of Learning*, 294 [VI.iii]).

72. Malthus, first *Essay*, 75.

73. Malthus, *Occasional Papers*, 45.

74. Malthus, first *Essay*, 139.

75. Malthus, first *Essay*, 139.

76. For a sample of the controversy, see Engels, "Outlines of a Critique of Political Economy," 222; Godwin, *Remarks*, 61; and James, *Population Malthus*, 400–401.

77. K. Smith, *Malthusian Controversy*, 210 and 234. Elsewhere, in a pamphlet on high prices, Malthus commented on his reason for evidencing extreme, implausible examples. "In the complicated machinery of human society, the effect of any particular principle frequently escapes from the view, even of an attentive observer, if it be not magnified by pushing it to extremity," he wrote (Malthus, *Pamphlets*, 18).

78. Kant, *Critique of Judgment*, 106.

79. Kant, *Critique of Judgment*, 128.

80. Malthus, second *Essay*, I, 3. See also Isaacson, *Malthusian Limit*, viii.

81. Ellul's second law of "self-augmentation" re-circulated this line of thought when he wrote, "Technical progress tends to act, not according to an arithmetic, but according to a geometric principle" (*Technological Society*, 89).

82. Malthus, first *Essay*, 74.

83. Malthus, first *Essay*, 74.

84. Malthus, first *Essay*, 74.

85. Malthus, first *Essay*, 138.

86. Malthus, first *Essay*, 138–39.

87. Malthus, second *Essay*, II, 127. The second *Essay*'s feast metaphor became one of its most infamous passages and garnered so much harsh criticism that Malthus excised this passage from the 1806 and later editions of the *Essay*. Its excision did not prevent future calumniation by rival political parties. For instance, the feast metaphor became a notorious protoexample of social Darwinism. See Proudhon, *Malthusians*, 6–7; and Rocker, *Nationalism and Culture*, 470.

88. Hill, "Rhetorical Transformation of the Masses."

89. Foucault, *"Society Must Be Defended,"* 247.

90. Hobbes, *Leviathan*, 70.

91. James, *Population Malthus*, 401.

92. Petty, "Another Essay in Political Arithmetick," 464; Petersen, *Malthus*, 173.

93. Engels, "Socialism: Utopian and Scientific," 690.

94. Bonar, *Malthus and His Work*, 266.

95. Malthus, first *Essay*, 75.

96. Young, "Malthus on Man," 73.

97. Malthus, first *Essay*, 80.

98. See Walzer, "Logic and Rhetoric in Malthus's 'Essay,'" 15.

99. Malthus, first *Essay*, 83.

100. Malthus, first *Essay*, 84.

101. Malthus, second *Essay*, I, 13.

102. See Mbembe, "Necropolitics," 22–30.

103. Malthus, second *Essay*, I, 13.

104. Malthus, first *Essay*, 125 and 225; Malthus, second *Essay*, I, 444.

105. Malthus, first *Essay*, 118–19.

106. Malthus, first *Essay*, 138–39.

107. See Malthus, second *Essay*, I, 16.

108. The Ukrainian forced famine of 1932 and 1933, or Holodomor, during which about five million people starved at the behest of Stalin, demonstrated how food can be used as a weapon. See Serbyn's "Holodomor—The Ukrainian Genocide."

109. Malthus, second *Essay*, II, 189.

110. Young, "Malthus on Man, 82; Marx, *Grundrisse*, 605; Walzer, "Logic and Rhetoric in Malthus's 'Essay.'"

111. Keynes, "Malthus and Biological Equilibria"; Petersen, *Malthus*, 218–30; Young, "Malthus on Man."

112. Lanham, *Handlist of Rhetorical Terms*, 16.

113. Hill, "Rhetorical Transformation of the Masses."

114. According to comments they made, both Malthus and Ricardo were aware of the possible confusion caused by *antistasis* and polysemy. In the preface to *Definitions in Political Economy*, Malthus noted that "the differences of opinion among political economists have of late been a frequent subject of complaint; and it must be allowed, that one of the principal causes of them may be traced to the different meanings in which the same terms have been used by different writers" (vii). Malthus's refusal to use standardized meanings irked Ricardo and caused contention among their colleagues (Ricardo, *Works and Correspondence*, vol. 8, 331; Bonar, *Malthus and His Work*, 418).

115. Glebkin, "Socio-Cultural History of the Machine Metaphor."

116. Hobbes, *Leviathan*, 5.

117. Godwin, *Enquiry Concerning Political Justice*, 576.

118. Godwin, *Enquiry Concerning Political Justice*, 576.

119. Godwin, *Enquiry Concerning Political Justice*, 576.

120. Malthus used the machine metaphor often, and one of the more remarkable instances of machine *antistasis* in his writings is noteworthy even if it is tangential to his political argument with Godwin. In quick succession in *An Inquiry into the Nature and Progress of Rent*, Malthus refers to factory machines and to the earth as a "vast machine," the soil as "a great number of machines," land as machines, and agricultural technologies as machines (*Pamphlets*, 206–8).

121. Malthus, first *Essay*, 144; second *Essay*, I, 325–26.

122. As if in corroboration of the confusion caused by Malthus's terminology and the machine metaphor, commentators have ascribed other "main-springs" to Malthus's conception of society. An anonymous critic dubbed the entire "principle of population" the "master-spring in this social and political machine," and Jacques Dupâquier contradicted this passage in Malthus to single out marriage as the "central pivot of the regulating mechanism" (quoted in James, *Population Malthus*, 111; and Dupâquier, *Malthus Past and Present*, xii).

123. Malthus, second *Essay*, II, 203. See also Malthus, first *Essay*, 144.

124. Foucault would call the simultaneous governmental administration of massed bodies and the self-care of individual bodies an instance of "biopower" in action (*History of Sexuality*, 139–42).

125. Malthus, first *Essay*, 179.

126. Malthus, *Principles of Political Economy*, 360.

127. See Booth, *Company We Keep*, 108 and 189.

128. Malthus, first *Essay*, 155.

129. In "Malthus's Total Population Theory," J. J. Spengler addressed the impact of the concept of perpetual technological "progress" on Malthus's formulation of the "principle of population" (58–61).

130. Malthus, first *Essay*, 103 and 250. See also Malthus, second *Essay*, II, 82.

131. Malthus, second *Essay*, II, 32–34; Malthus, *Principles of Political Economy*, 352.

132. Malthus, *Principles of Political Economy*, 352; Malthus, *Pamphlets*, 118; G. Gilbert, "Economic Growth and the Poor," 191.

133. Richards, *Philosophy of Rhetoric*, 13; Greene, *Malthusian Worlds*, 8.

134. L. Marx, "Technology."

135. Malthus, second *Essay*, II, 185.

136. Malthus, first *Essay*, 210.

137. Malthus, second *Essay*, I, 3.

138. Farrell, "Sizing Things Up," 7.

139. "The Malthusian League" was a British contraception advocacy group that operated in the late nineteenth and early twentieth centuries. On Malthusian and anti-Malthusian abortion rhetoric, see Stormer's *Sign of Pathology*.

140. O'Mathúna, "Human Dignity in the Nazi Era"; Meek, *Marx and Engels on the Population Bomb*, 48–49.

141. Greene, *Malthusian Worlds*.

142. Greene, *Malthusian Worlds*, 15.

Chapter 2

The epigraphs for this chapter are drawn from Spies, *Autobiography*, 6, and Gary, "Chicago Anarchists," 831.

1. This historical account of Haymarket is derived from the Haymarket Affair Digital Collection (hereafter, HADC); Avrich, *Haymarket Tragedy*; David, *History of the Haymarket Affair*; Green, *Death in the Haymarket*; Nelson, *Beyond the Martyrs*; Rosemont and Roediger, *Haymarket Scrapbook*; and Schaack, *Anarchy and Anarchists*.

2. On Spies's involvement in the political, social, and cultural context of Chicago's German-speaking working class neighborhoods, see "Aurora Turnverein," 165–68; Heiss, "German Radicals in Industrial Chicago," 208, 210, and 213; Heiss, "Popular and Working-Class German Theater in Chicago," 194; Keil, "German Immigrant Working Class of Chicago," 166–67; Keil and Ickstadt, "Elements of German Working-Class Culture in Chicago, 97; Kruger, "Cold Chicago," 21; Nelson, *Beyond the Martyrs*, 131 and 267; and Poore, *German-American Socialist Literature*, 103.

3. Issued by the Socialist Publishing Society, the paper was collectively edited, but Spies was the unofficial editor-in-chief. Readership reached twenty thousand by the mid-1880s (Nelson, "*Arbeitspresse und Arbeiterbewegung*"). For more on the *Arbeiter-Zeitung* and the anarchist press, see Bekken, "First Anarchist Daily Newspaper"; Keil, "Introduction," xviii; Keil and Ickstadt, "Elements of German Working-Class Culture"; Keil and Jentz, *German Workers in Chicago*, 242–43 and 300; Kiesewetter, "German-American Labor Press"; and Nelson, *Beyond the Martyrs*, 115–26.

4. HADC, People's Exhibit 63.

5. Nietzsche, *Friedrich Nietzsche on Rhetoric and Language*, 250.

6. HADC, Testimony of Edgar E. Owen.

7. HADC, Testimony of Albert Parsons, 126.

8. HADC, Testimony of Albert Parsons, 124.

9. Black, Salomon, and Zeisler, *Brief and Argument*, 122.

10. HADC, Testimony of James Bowler.

11. McLean, *Rise and Fall of Anarchy in America*, 18–19.

12. A czar bomb, or globe bomb, is the stereotypical dynamite bomb of the era, consisting of a round, iron, ball-shaped casing with a wick protruding from the top. See HADC, "People's Exhibit 129A"; and HADC, Testimony of Harry Wilkinson. Incendiary was a common label attached to labor agitation by the authorities. The Haymarket police report, for instance, described the evening's speeches as incendiary (HADC, "Inspector John Bonfield report to Frederick Ebersold," 4 and 6).

13. Spies, "Address of August Spies," 5; "Hellish Deed."

14. Floyd Dell, a Chicago literary critic, poet, and radical journalist, called this type of dynamite discourse "bomb-talking" ("Socialism and Anarchism in Chicago," 388–92). The HADC's "People's Exhibits" provide many examples of preaching dynamite. For more on the prevalence of dynamite rhetoric in the Haymarket era, see Adamic, *Dynamite*; Avrich, *Haymarket Tragedy*, 160–77; Barnard, *Eagle Forgotten*, 80–82; Gary, "Chicago Anarchists," 813–19; Larabee, *Wrong Hands*, 15–35; and Nelson, *Beyond the Martyrs*, 161–63.

15. Spies, "Address of August Spies."

16. Johann Joseph Most's *Revolutionäre Kriegswissenschaft* was the most notorious piece of anarchist dynamite advocacy at the Haymarket trial (Larabee, *Wrong Hands*, 18).

17. Gary, "Chicago Anarchists," 831.

18. "Capt. Schaack's Day"; Black, Salomon, and Zeisler, *Brief and Argument*, 164–73.

19. "Capt. Schaack's Day."

20. Black, Salomon, and Zeisler, *Brief and Argument*, 164–73.

21. On objects and their systemic capacity for communication, see Feigenbaum's "Resistant Matters" (16). On the ways that "evidence speaks for itself" in law, see Daston's "Introduction: Speechless" (12–13).

22. Perelman and Olbrechts-Tyteca, *New Rhetoric*, 117.

23. On the materiality of pathos, see Barnett's "Chiasms."

24. "Hellish Deed."

25. See Moore, "Life, Liberty, and the Handgun," 437; and McGee, "'Ideograph,'" 9. "REVENGE!" was added by Hermann Pudewa, an *Arbeiter-Zeitung* typesetter.

26. Moore, "Life, Liberty, and the Handgun," 436.

27. HADC, People's Exhibit 6.

28. HADC, Testimony of August Spies, 100.

29. McGee, "'Ideograph,'" 4.

30. Spies, *Autobiography*, 6.

31. May, "Orator-Machine," 440.

32. May, "Orator-Machine," 440.

33. See Avrich, *Haymarket Tragedy*, 72–75; and Nelson, *Beyond the Martyrs*, 153–73.

34. Schaack, *Anarchy and Anarchists*, 390.

35. Foner, *Autobiographies of the Haymarket Martyrs*, 42–43.

36. Bennett, *Vibrant Matter*, xvi and xii.

37. Farrell, *Norms of Rhetorical Culture*, 63.

38. Browne, "'This Unparalleled Inhuman Massacre,'" 311.

39. Ceccarelli, "Polysemy," 398.

40. Ceccarelli, "Polysemy," 396–97. Ceccarelli called this type of polysemy "strategic ambiguity," but I prefer the term "strategic polysemy," a term that Ceccarelli used in passing (397) rather than as part of her primary critical terminology.

41. Harding, *Sciences from Below*, 182.
42. Richards, *Philosophy of Rhetoric*, 11.
43. *Chicago Tribune*, May 5, 1886.
44. In *Urban Disorder*, Carl Smith called the unstable, ironic meanings and contested definitions of words—such as "system," "civilization," and "anarchism," which were prevalent during the Haymarket events—"ironic reversals," a term he also used to describe the hurling of accusations back against accusers (143–46).
45. Spies, *Reminiszenzen von August Spies*, 4, translation mine; Spies, "Address of August Spies," 4.
46. Spies, "Address of August Spies," 5.
47. In this period, "dynamite" became a generic term for Nobel's mixture, a mixture that people could imitate with ease by substituting other absorbent materials, most often sawdust, for kieselguhr (Bown, *Most Damnable Invention*, 82).
48. Spies, "Address of August Spies," 3; Parsons, "Address of Albert R. Parsons," 82; Avrich, *Haymarket Tragedy*, 69 and 176.
49. Most, *Revolutionäre Kriegswissenschaft*, 3–4 and 15.
50. Bakunin, *God and the State*, 40, 59, and 62; Bakunin, *Statism and Anarchy*, 159; Kropotkin, "Modern Science and Anarchism," 150–52, 168, and 172.
51. Spies, "Address of August Spies," 15.
52. Spies, "Address of August Spies," 15.
53. Spies, "Address of August Spies," 15.
54. Spies, "Address of August Spies," 3.
55. "Infernal Machines." Much of the dynamite reportage in the *Arbeiter-Zeitung* was republished from mainstream press articles about military affairs. Even the *Encyclopedia Britannica* detailed how to manufacture dynamite bombs.
56. Spies, "Address of August Spies," 9.
57. Spies, "Address of August Spies," 9.
58. Spies, "Address of August Spies," 6.
59. Spies, *Autobiography*, 38, emphasis in original.
60. Spies, "Address of August Spies," 6 and 12.
61. Spies, "Address of August Spies," 6.
62. In *Shakespeare's Use of the Arts of Language*, Sister Miriam Joseph defined *antistrephon* as "a captious argument which turns that which serves the opponent's purpose to one's own" (199). Joseph called *antistrephon* "captious" in reference to Thomas Wilson's classification of "trappying argumentes," so called "because fewe that aunswere vnto theim, can auoide daunger" (*Rule of Reason*, 210). Wilson defined the trapping argument *antistrephon* as "nothing els, then to tourne a mannes saiying into his awne necke again, and to make that whiche he bringeth for his awne purpose, to serue for our purpose" (*Rule of Reason*, 211). Richard Lanham defined it as "an argument that turns one's opponent's arguments or proofs to one's own purpose," differentiating *antistrephon* from *metastasis*, which refers to turning insults around to insult the initial insulter (*Handlist of Rhetorical Terms*, 16 and 101).
63. Spies, "Address of August Spies," 1, emphasis in original.
64. Spies, "Address of August Spies," 1 and 8.
65. Spies, "Address of August Spies," 9.
66. Spies, "Address of August Spies," 9.
67. Spies, *Autobiography*, 32–33.
68. Spies, "Address of August Spies," 9.
69. Spies, "Address of August Spies," 14 and 16.
70. Spies, "Address of August Spies," 5.
71. Spies, "Address of August Spies," 14.

72. Spies, "Address of August Spies," 2–3.

73. Spies, "Address of August Spies," 5.

74. Spies, "Address of August Spies," 13.

75. Spies, "Address of August Spies," 8–9.

76. Spies, "Address of August Spies," 7–8.

77. Spies, "Address of August Spies," 3.

78. Griffin, "'Operation Sunshine,'" 522.

79. As Jeffory A. Clymer wrote, preaching dynamite empowered agitators to "gauge the possibilities for—and limits on—social action and individual agency in the emerging mass culture produced by industrial and finance capitalism" (*America's Culture of Terrorism*, 24).

80. Quoted in Avrich, *Haymarket Tragedy*, 163–64; Most, *Revolutionäre Kriegswissenschaft*, 10 and 21.

81. HADC, People's Exhibit 21.

82. Exaggerations of weapons' capacities for destruction have proven self-defeating for dissenting against weapons. See Hogan, *Nuclear Freeze Campaign*; and Ivie, "Metaphor," 168.

83. For accounts of the public's fear of dynamite after Haymarket, see Avrich, *Haymarket Tragedy*, 177 and 222–24; Green, *Death in the Haymarket*, 199–200; C. Smith, *Urban Disorder*, 124–25; and the earwitness account of Samuel P. McConnell, "Chicago Bomb Case," 730.

84. Quoted in Avrich, *Haymarket Tragedy*, 455–56.

85. Quoted in Avrich, *Haymarket Tragedy*, 454–55.

86. Nelson, "*Arbeitspresse und Arbeiterbewegung*," 103.

87. Kropotkin, "Anarchism," 916. See also Rosemont's "Bomb-Toting, Long-Haired, Wild-Eyed Fiend."

88. Solomon, "Ideology as Rhetorical Constraint," 185.

89. Green, *Death in the Haymarket*, 203.

90. Clymer, *America's Culture of Terrorism*, 52 and 61.

91. Clymer, *America's Culture of Terrorism*, 24.

92. Altgeld, "Altgeld's Reasons," 153.

93. "Hellish Deed."

94. Quoted in Green, *Death in the Haymarket*, 192–93.

95. Spies, *Autobiography*, 25.

96. Quoted in Avrich, *Haymarket Tragedy*, 455.

97. Rosemont, "Bomb-Toting, Long-Haired, Wild-Eyed Fiend."

98. Spies, "Address of August Spies," 16.

99. For overviews of the Haymarket executions' influence on labor movements throughout the world, see Avrich, *Haymarket Tragedy*, 428–36; Green, *Death in the Haymarket*, 274–320; and Rosemont and Roediger, *Haymarket Scrapbook*, 173–248.

100. Spies, "Address of August Spies," 16.

101. Burke, *Grammar of Motives*, 516. Burke wrote of the "synecdochic representative."

102. Clymer, *America's Culture of Terrorism*, 41.

103. Spies, "Address of August Spies," 2.

104. Browne, *Angelina Grimké*, 37. Writing about the abolitionist movement to end slavery in the United States, Browne contended that, as an "object of interpretation," violence was "constitutive of discourse itself" (36–37).

105. Louis Adamic metonymically titled his history of American labor conflict *Dynamite: The Story of Class Violence in America*. See also Gage, *Day Wall Street Exploded*; Grover, *Debaters and Dynamiters*; and T. Jones, *More Powerful Than Dynamite*.

Chapter 3

The epigraphs for this chapter are drawn from Fries and West, *Chemical Warfare*, 370, and Irwin, *Next War*, 44.

1. For biographical information about Fries, see the brief biography that accompanies his article "Chemical Warfare," the article "Brigadier General Amos A. Fries," and Fries's January 1920 testimony for the House of Representatives hearings on War Expenditures. Thomas Faith's history of the Chemical Warfare Service (CWS), *Behind the Gas Mask*, is the most detailed academic account of Fries's activities with the CWS.

2. Fries, "Road Work by the Army"; Fries, "Los Angeles Harbor"; Fries, "San Pedro Harbor."

3. Mumford, *Myth of the Machine*, 55; McNeill, *Pursuit of Power*, 308 and 314. Preston's *Higher Form of Killing* provides a useful primer of World War I weapons development.

4. McNeill, *Pursuit of Power*, 310–17.

5. McNeill, *Pursuit of Power*, 316–45.

6. L. F. Haber, *Poisonous Cloud*, 192. My historical rendering of chemical warfare and the CWS in World War I and after is drawn from Haber's *The Poisonous Cloud*, the three-volume official military history of chemical warfare during and between World War I and World War II (Brophy and Fisher, *Chemical Warfare Service: Organizing for War*; Brophy, Miles, and Cochrane, *Chemical Warfare Service*; Kleber and Birdsell, *Chemical Warfare Service*), the US War Department's journal *Chemical Warfare Service*, Faith's *Behind the Gas Mask*, and Fries's writings.

7. Whitehead, "Third Ypres," 195.

8. Manley, "Problem of Old Chemical Weapons," 2.

9. Brophy, Cochrane, and Miles, *Chemical Warfare Service*, 2–5; Fries and West, *Chemical Warfare*, 72; Heller, *Chemical Warfare in World War 1*, 47.

10. Fries to Chamberlain, September 16, 1919, Amos A. Fries Papers, Box 15 A–C, Chamberlain.

11. Sibert, *Report of the Chemical Warfare Service 1918*, 3. This document and Sibert's *Report of the Chemical Warfare Service 1919* detailed the CWS's organization and activities.

12. As a pundit, Fries often testified before Congress on educational matters, and in 1938, Fries became the editor of the "right wing" anticommunist educational bulletin called *Friends of the Public Schools of America*, a position he held until 1953 (see the University of Iowa's "Right Wing Collection").

Fries devoted a significant amount of space to military and other political issues in the education bulletin. The publication took a negative position on the 1948 and 1951 Genocide Conventions, the 1950 Stockholm Peace Petition that sought to outlaw nuclear war, and the 1948 Universal Declaration of Human Rights, arguing that these proposals amounted to the first phase of implementing a world government that would destroy hard-won American freedoms (Fries, "Genocide Convention," 4; Fries, "Stockholm Peace Petition," 1–3; Fries, "Genocide Convention Side-Tracked for Human Rights Declaration," 5–6).

Aside from material reprinted from other newspapers and serials, almost all of the articles and opinion pieces in *Friends* are anonymous. However, I suspect that many of them are written by Fries, because they have the same distinctive typographic and editorial nuances (e.g., many boldfaced and capitalized passages) that Fries used throughout both *Sugar Coating Communism for Protestant Churches* and *Communism Unmasked*.

13. Allen, "Chemical Warfare," 33.

14. Fries and West, *Chemical Warfare*, 90.

15. Fries to Sibert, October 15, 1919, Amos A. Fries Papers, Box 15 A–C, Chief, Chemical Warfare Service. On occasion Fries and his interlocutors referred to their pro-gas activities as "propaganda" (Schulz to Fries, August 11, 1919 and Fries to Schulz, August 15, 1919, Amos A. Fries Papers, Box 19 O–S, Col. J. W. N. Schulz; Pope to Fries January 29, 1919, Amos A. Fries Papers, Box 19 O–S, Maj. F. Pope).

16. Fries to Atkisson, October 30, 1919, Amos A. Fries Papers, Box 15 A–C, Col. Atkisson; Fries to Anderson, October 15, 1919, Amos A. Fries Papers, Box 15 A–C, Maj. W. Anderson; Fries to Chadbourne, October 3, 1919 and Chadbourne to Fries, October 27, 1919, Amos A. Fries Papers, Box 15 A–C, Chadbourne; Fries to Clark, October 11, 1919, Amos A. Fries Papers, Box 16 C–E, Col. E. B. Clark; Fries to Law, August 16, 1919, Box 17 E–L, Capt. J. D. Law; Fries to McKenzie, October 11, 1919, Amos A. Fries Papers, Box 18 L–O, Hon. J. C. McKenzie.

17. Amos A. Fries Papers, Box 19 O–S, Resolutions (also reproduced in F. Brown, *Chemical Warfare*, 79); Vilensky, *Dew of Death*, 26; D. Jones, "From Military to Civilian Technology," 159–60; Fries, review of *Chemical Warfare*.

18. Fries to Atkisson, September 25, 1919, Amos A. Fries Papers, Box 15 A–C, Col. Atkisson.

19. Faith, *Behind the Gas* Mask, 98–100.

20. Allen, "Chemical Warfare," 33. See also F. Brown, *Chemical Warfare*, 130; Russell, *War and Nature*, 39; Fries, "Letters from America," 14.

21. In *Chemical Warfare*, the two authors quoted wholesale from papers found in chemistry journals and army files, and Fries and West acknowledged their rampant borrowing in the book's preface (vii). Much of *Chemical Warfare* is a compilation of papers written separately by either Fries or West, so their dual authorship appears to be a result of publishing separately written papers under the same cover, rather than a result of collaborative writing.

Regarding Fries's authorship, Fries figured that West would probably end up authoring more of the book (Fries to West, October 21, 1919, Amos A. Fries Papers, Box 20 S–Z, Maj. C. J. West). Fries was responsible for covering "the Field side" or "tactical side" of chemical warfare (West to Fries, August 11, 1919, and Fries to West, August 14, 1919, Amos A. Fries Papers, Box 20 S–Z, Maj. C. J. West). "Gas in Defense," a paper held in the Munitions Supply Branch Library, is a draft of *Chemical Warfare*'s chapter "Defense against Gas." "Chemical Warfare in Attack" is "The Offensive Use of Gas." "Gas in Attack and Defense" is a draft of the book's chapters on offensive and defensive chemical warfare and duplicates both "Gas in Defense" and "Gas in Attack." According to a footnote, Fries addressed students of the General Staff College in Washington, DC, on May 11, 1921, which became the basis for the *Chemical Warfare* chapter titled "Chemical Warfare in Relation to Strategy and Tactics" (363). "A History of Chemical Warfare in France," a paper held at the Edgewood Arsenal Historical Office, is probably a draft of *Chemical Warfare*'s "The Chemical Warfare Service in France," but I have been unable to verify this.

Regarding West's authorship, the project, first conceived of as a "history of the Chemical Warfare Service" or "a semi-official history of Chemical Warfare," was West's idea, and he was to be in charge of covering "the chemical side" of warfare (West to Fries, August 11, 1919 and West to Fries, August 21, 1919, Amos A. Fries Papers, Box 20 S–Z, Maj. C. J. West). *Chemical Warfare*'s first chapter, "The History of Poison Gases," is a reprint of an article first published in *Science* and he also wrote chapters 2 and 3 (West to Fries, October 15, 1919, Amos A. Fries Papers, Box 20 S–Z, Maj. C. J. West). West probably wrote all or the bulk of chapters 12–15 of *Chemical Warfare* about gas masks, absorbents, and other defensive measures, since gas masks and canisters were the focus of his research throughout 1918–19. West also wrote "The Chemical Warfare Service," a general survey of the CWS's contributions to science.

22. Brophy, Miles, and Cochrane, *Chemical Warfare Service*, 25. See Col. George A. Burrell's "Research Division, Chemical Warfare Service, U.S.A." for an overview of the research division's wartime activities.

23. Fries, "Summary of Marine Piling Investigation."

24. West's more time-consuming postwar work was as a scientific bibliographer and librarian of sorts, first at the NRC, as director of the Information Department; then later at Arthur D. Little, Inc., a paper-research company responsible for products such as gas masks' "sucked-on paper" (Logan to Fries, May 7, 1920, Amos A. Fries Papers, Box 15 A–C, Chief, Chemical Warfare Service); and finally as a research associate at the Institute of Paper Chemistry at Lawrence College. He chronicled US science with dozens of NRC bulletins and bibliographies on topics as diverse as molasses, metallurgy, and scientific doctorates conferred by American universities.

25. Parsons, "American Chemist in Warfare," 780.

26. Fries to Sibert, July 17, 1919, Amos A. Fries Papers, Box 15 A–C, Chief, Chemical Warfare Service.

27. Carl W. Ackerman's introduction to Fries, "Future of Poison Gas," 419.

28. J. Scott, *Hague Conventions*, 116 and 225.

29. *Conference on the Limitation of Armament*, 25. This assertion was an oft-repeated commonplace of chemical-weapons discourse. See Faith, *Behind the Gas* Mask, 83 and 104.

30. Quoted in Russell, *War and Nature*, 5. This passage appears often in chemical-warfare advocacy, and Fries and West reproduced it in *Chemical Warfare* (6).

31. The issue of weapons' "humanity" and "humaneness" had been an element of weapons diplomacy since at least the 1860s. See "Declaration of St. Petersburg, 1868"; and Fries and West, *Chemical Warfare*, 5.

32. Haldane, *Callinicus*, 52–53.

33. Although "inhuman" and "inhumane" (and "human" and "humane") connote different meanings, chemical-warfare discourse tended to treat these terms as synonymous. Fries and West, for instance, posed "humane" versus "inhuman." To rank the two words in terms of their negative connotations, "inhuman" perhaps ranks as more derogatory than "inhumane." "Inhuman" connotes that its object is so despicable that it is beyond the scope of how humanity should be defined. "Inhumane" connotes that its object is at least "human," but its object is tainted by a failure to adhere to ethical and moral codes. I have not found any indication that any participants in chemical-warfare discourse intended to connote one level of negative or positive description rather than another by using these terms. Thus, my analysis treats the two terms as synonymous.

34. Burke, *Permanence and Change*, 257.

35. Burke, *Permanence and Change*, 255.

36. Burke, *Attitudes toward History*, 47. See also Prelli, Anderson, and Althouse's "Kenneth Burke on Recalcitrance."

37. Mitchell, *Strategic Deception*.

38. Hook, *Fail-Safe Fallacy*.

39. S. Fuller, *Philosophy, Rhetoric, and the End of Knowledge*, 244.

40. Richter, *Chemical Soldiers*, 1.

41. Burke, *Permanence and Change*, 258.

42. Burke, *Permanence and Change*, 258.

43. Jones, "From Military to Civilian Technology"; Russell, *War and Nature*.

44. Winner, "Upon Opening the Black Box," 372.

45. Farrell, "Sizing Things Up," 1.

46. Farrell, "Sizing Things Up," 1.

47. Kuhn, "Rhetoric and Liberation," 9–11; Kuhn, *Structure of Scientific Revolutions*, 94.

48. See Farrell, "Knowledge, Consensus, and Rhetorical Theory." The University of Iowa's Project on Rhetoric of Inquiry (POROI) seeks to study the rhetoric of epistemology: "In its most general sense, the rhetoric of inquiry attempts to apply the tools of rhetorical criticism, whose native soil is the critique of political speech, to the production, dissemination, and reception of what counts as knowledge. This research program runs against modernity's conventional opposition between truth and persuasion" (Depew, "Introduction," 1).

49. Schiappa, *Defining Reality*.

50. Fries and West, *Chemical Warfare*, 176.

51. See Warthin and Weller's *Medical Aspects of Mustard Gas Poisoning*, published in 1919, which drew from forty-nine already-published articles on the topic.

52. Reproduced in M. Gilbert, *First World War*, 352–53. Owen's description seems to indicate an attack with multiple types of gas—mustard, as evidenced by the relative silence of the shells and the blisters, and perhaps phosgene or chlorine, based on the descriptions of choking and drowning.

53. L. F. Haber, *Poisonous Cloud*, 189–90.

54. Fries and West, *Chemical Warfare*, 177.

55. Brice, *Battle Book of Ypres*, 36–37.

56. Hessel, Martin, and Hessel, *Chemistry in Warfare*, 93; Brice, *Battle Book of Ypres*, 182.

57. Auld, *Gas and Flame*, 178.

58. Steel and Hart, *Passchendaele*, 279; MacDonald, *They Called It Passchendaele*, 87.

59. MacDonald, *They Called It Passchendaele*, 87–88.

60. "Drops Balls of Gas on American Lines."

61. F. Brown, *Chemical Warfare*, 37.

62. Warthin and Weller, *Medical Aspects of Mustard Gas Poisoning*, 20.

63. Fries and West, *Chemical Warfare*, 426.

64. Allen, *Toward the Flame*, 92.

65. Allen, *Toward the Flame*, 93.

66. F. Brown, *Chemical Warfare*, 36.

67. Jones and Wessely, *Shell Shock to PTSD*, 40; E. Brown, "Between Cowardice and Insanity."

68. Fries and West, *Chemical Warfare*, 175. See also Brophy, Miles, and Cochrane, *Chemical Warfare Service*, 62; Hessel, Martin, and Hessel, *Chemistry in Warfare*, 92; and Manning, *War Gas Investigations*, 16.

69. Fries and West, *Chemical Warfare*, 89.

70. "Submarines and Gas Condemned by Public;" quoted in F. Brown, *Chemical Warfare*, 69.

71. Faith, *Behind the Gas* Mask, 84.

72. See the front matter in Fries and West, *Chemical Warfare*.

73. Kleber and Birdsell, *Chemical Warfare Service*, 653.

74. Irwin, *Next War*, 44.

75. Quoted in Fries, *Excerpts from the Annual Report* (1922), 278.

76. Ogden and Richards, *Meaning of Meaning*, 12.

77. See Hill, "Memes, Munitions, and Collective Copia."

78. Fries and West, *Chemical Warfare*, 89.

79. The CWS's demobilization order cut its equipment allocations to zero and officially cancelled all outstanding contracts (Harrison to Sibert, November 29, 1918, Amos A. Fries Papers, Box 15 A–C, Chief, Chemical Warfare Service). In the end, armistice brought the ninety-seven-percent demobilization of the CWS and millions of dollars of unfilled chemical-production contracts (Brophy, Miles, and Cochrane, *Chemical Warfare Service*, 24; Fries, *Excerpts from the Annual Report* [1921], 222).

80. Fries to Harbord, June 10, 1919, Amos A. Fries Papers, Box 16 C–E, Chief of staff, AEF Paris.

81. Fries and West, *Chemical Warfare*, 387 and 370.

82. The same rhetorical tactics, as well as many sentences identical to those in *Chemical Warfare* that tout the humanity of gas, appear in Fries's 1919 article "The Humanity of Poisonous Gas."

83. Fries understood the persuasive force of a well-placed and manipulated statistical argument, as evidenced by a Memorial Day address he delivered in 1931. "There is an old and very true saying that 'Figures won't lie, but liars will figure. . . . I say this just before I quote you some figures,'" he said (*Memorial Day, 1931*, 3).

84. Fries and West, *Chemical Warfare*, 388. On September 10, 1919, Fries had written a draft of this argument titled "The Measure of Inhumanity in any Method of Fighting" (Amos A. Fries Papers, Box 19 O–S, no folder title). For similar arguments that use casualty and fatality statistics to claim that chemical warfare is not inhuman, see Colonel C. F. Brigham's (of the Chemical Warfare Service) and W. Lee Lewis's (the inventor of Lewisite) arguments in Villard, "Poison Gas," 32; Crowell and Wilson, *Armies of Industry*, 490–91; Francine, "Is Chemical Warfare More Inhuman Than Gunfire?" 2; Haldane, *Callinicus*, 27 and 32–33; Gilchrist, *Comparative Study of World War Casualties*, 46–50; Waitt, *Gas Warfare*, 5; US Bureau of Naval Personnel, *Gas! Know Your Chemical Warfare*, 9; Hammond, *Poison Gas*, 35; Mauroni, *America's Struggle with Chemical-Biological Warfare*, 5–6; and Sibert, *1918*, 17.

85. Fries and West, *Chemical Warfare*, 388. See also pp. 13 and 370. In his response to "Casualties from Gas in the A. E. F. (Official Figures Compiled by Surgeon General's Office)," Fries had already drafted this statistical argument (Amos A. Fries Papers, Box 19 O–S, no folder title). The difference between the 27.3 percent and 27.4 percent rates might be attributable to their use of different sources, or they may have been referring to different types of casualties (e.g., American casualties versus total war casualties). For a different report from the Surgeon General of the AEF on gas casualty statistics, see "Gas Casualties in A. E. F. Compiled from Gas Officers' Reports," Amos A. Fries Papers, Box 15 A–C, AEF Misc.

86. Fries and West, *Chemical Warfare*, 386.

87. In 1918, Sibert made the statistical humaneness of chemical weapons sound almost scientific: "Gas is twelve times as humane as bullets and high explosives," he asserted (*Report of the Chemical Warfare Service 1918*, 17).

88. Fries, *Sugar Coating Communism*, 62.

89. Clausewitz, *On War*, 83–85.

90. Fries and West, *Chemical Warfare*, 169.

91. On the primacy of the CWS's pedagogical onus, see Sibert's foreword to Fries and West's *Chemical Warfare* (x); Auld, *Gas and Flame*, v and 169; Haldane, *Callinicus*, 72; and Hammond, *Poison Gas*, ix.

92. Fries and West, *Chemical Warfare*, 374, emphasis in original.

93. Fries and West, *Chemical Warfare*, 65.

94. Fries and West, *Chemical Warfare*, 414.

95. Fries and West, *Chemical Warfare*, vii.

96. Fries and West, *Chemical Warfare*, 171.

97. Fries and West, *Chemical Warfare*, 417.

98. The "lethal concentration" of mustard gas is "30-minute exposure to 0.07 oz./1,000 cu. ft. air (when breathed)" (US War Department, *Technical Manual No. 3–215*, 122). A "minimum effective concentration" of mustard gas that produces casualties is one part mustard gas to twelve million five hundred thousand parts of air (Manning, *War Gas Investigations*, 26).

99. Fries and West, *Chemical Warfare*, 86.

100. Fries and West, *Chemical Warfare*, 86–87.

101. Holley, *Ideas and Weapons*, 162. For a critique of World War I's casualty and fatality statistics, see Faith, *Behind the Gas Mask*, 66–67.

102. Fries to Sibert, October 10, 1919, Amos A. Fries Papers, Box 15 A–C, Chief, Chemical Warfare Service.

103. Fries and West, *Chemical Warfare*, 150 and 176.

104. Fries and West, *Chemical Warfare*, 169. See also pp. 178–79 and 437.

105. Fries and West, *Chemical Warfare*, 437.

106. Fries and West, *Chemical Warfare*, 178. See also Fries, "Address," 2.

107. Fries and West, *Chemical Warfare*, 371.

108. Fries, "Chemical Warfare," 424.

109. Fries, "Chemical Warfare," 436.

110. Fries, *Excerpts from the Annual Report* (1921), 226; Fries, "Future of Poison Gas," 421.

111. Weaver, *Ethics of Rhetoric*, 201.

112. Fries, *Communism Unmasked*, 196.

113. Fries, *Sugar Coating Communism*, 12, 25, and 39, emphasis in original.

114. Fries, *Sugar Coating Communism*, 39 and 49–50; Fries *Communism Unmasked*, 190, emphasis in original.

115. Fries *Communism Unmasked*, 188, emphasis in original.

116. Fries *Communism Unmasked*, 189, emphasis in original.

117. On at least one occasion, Fries advocated the use of tear gas to control rioting mobs as humane by mailing a one-page article, "Mob Control," along with "The Humanity of Poisonous Gas" to Norfolk, Virginia's Director of Public Safety, Rear Admiral A. C. Dillingham ("Mob Control," Amos A. Fries Papers, Box 15 A–C, Burnside; Fries to Dillingham, September 23, 1919, Amos A. Fries Papers, Box 16 C–E, Rear Admiral A. C. Dillingham).

118. See Hill, "Memes, Munitions, and Collective Copia."

119. Fries, "Uses and Dangers of Poisonous Gases," 55.

120. See Hill, "Memes, Munitions, and Collective Copia"; and Kant, *Perpetual Peace*.

121. Fries and West, *Chemical Warfare*, 383. On the "universality" of chemical weapons, see Fries, "Future of Poison Gas," 421; Fries, "Deadly Weapons of Chemical Warfare," 435; Fries, *Excerpts from the Annual Report* (1924), 12; and Fries and West, *Chemical Warfare*, 436.

122. Fries, *Excerpts from the Annual Report* (1922), 282.

123. Hughes, "Technological Momentum," 110.

124. See Fries and West, *Chemical Warfare*, 427–34. See also Fries's "By-Products of Chemical Warfare."

125. Fries and West, *Chemical Warfare*, 427.

126. Fries and West, *Chemical Warfare*, 427, 433, and 436.

127. See Russell, *War and Nature*.

128. Fries and West, *Chemical Warfare*, 169.

129. Fries and West, *Chemical Warfare*, 438.

130. Rappert, *Controlling the Weapons of War*, 4–5.

131. Fries and West, *Chemical Warfare*, 414–15.

132. Quoted in Brophy and Fisher, *Chemical Warfare Service*, 22.

133. Ellis, *Social History of the Machine Gun*, 52; Holley, *Ideas and Weapons*.

134. See US War Department, *Chemical Warfare Service Field Manual*. On the idea of the "comprehensive military doctrine," see Holley, *Ideas and Weapons*, 167.

135. See Faith, *Behind the Gas Mask*.

136. Freedman, *Deterrence*, 32.

137. Manley, "Problem of Old Chemical Weapons," 6–7.

138. Goodwin, *Keen as Mustard.*

139. Pechura and Rall, *Veterans at Risk.*

140. Sutherland, "Thiodiglycol," 28; Robinson and Trapp, "Production and Chemistry of Mustard Gas," 6; Kaplan, et al., "Summary and Conclusions," 127.

141. Sutherland, "Thiodiglycol," 24.

142. Fries, *Memorial Day, 1931.*

143. Fries, "Uses and Dangers of Poisonous Gases," 56.

144. Jones, "From Military to Civilian Technology," 164–65.

145. On the similarities and dissimilarities of these types of weapons manuals, see Larabee, *Wrong Hands*, 5.

146. Fries, "Uses and Dangers of Poisonous Gases," 56.

147. Even article 5 of 1948's Universal Declaration of Human Rights, still a definitive statement on international relations, used the "inhuman" stasis point to declare that "no one shall be subjected to torture or to cruel, inhuman or degrading treatment or punishment."

148. Preston, *Higher Form of Killing*, 5.

149. Kerry, "Statement on Syria."

150. Fries and West, *Chemical Warfare*, 438–39.

Chapter 4

The epigraphs for this chapter are drawn from Szilard, *His Version of the Facts*, vi, and Merton, *Sociology of Science*, 261.

1. My historical and biographical narrative about Szilard is derived from Szilard's reminiscences in *His Version of the Facts*, Richard Rhodes's *Making of the Atomic Bomb*, and three biographies of Szilard—Lanouette's *Genius in the Shadows*, Gandy's *Leo Szilard*, and Esterer and Esterer's *Prophet of the Atomic Age.*

2. Rhodes, *Making of the Atomic Bomb*, 27.

3. Wells, *World Set Free*, 152.

4. Wells, *World Set Free*, 153.

5. Szilard, *His Version of the Facts*, 17.

6. Gandy, *Leo Szilard*, xiii. See also Bethe, quoted in Lanouette, *Genius in the Shadows*, xix and 384; and Bess, *Realism, Utopia, and the Mushroom Cloud*, 53 and 56.

7. Perhaps no other invention exemplifies the presence of Technē's Paradox in weapons rhetoric better than what are known as atomic, nuclear, and hydrogen bombs. For this reason, I refer to the full range of nuclear bombs collectively as "the Bomb." "Names, as titles for situations," according to Charles Kauffman, "describe and define the conditions under which weapons can be used," "synecdochically . . . sum up under a single heading all the characteristics of the object," and grant weapons agential power ("Names and Weapons," 274–76).

8. Quintilian, *Institutio Oratoria of Quintilian*, 355 [XII.1.1].

9. Bernstein, "Dazzling Gadfly."

10. Szilard to Niels Bohr, November 7, 1950, Leo Szilard Papers.

11. Szilard helped Byron Miller and Edward Levi to draft the bill.

12. Szilard, *His Version of the Facts*, 94–96. Many of the proposals for how to organize research into atomic energy that he outlined in an August 1940 memorandum to Alexander Sachs were realized as the Manhattan Engineering District took shape (137–39). Not all atomic historians share the belief that Szilard's prodding influenced the United States' pursuit of the Bomb, because as a foreigner, the Briggs Committee excluded Szilard to an extent (Bundy, *Danger and Survival*, 36; A. Smith, *Peril and a Hope*, 27).

13. The US government awarded the patent in 1955. The "Metallurgical Laboratory" was "a code name chosen to conceal the nature of the work being done there" (Groves, *Now It Can Be Told*, 9).

14. On the surveillance of Szilard during and after the war, see Gruber's "Manhattan Project Maverick" (79–81 and 86–87).

15. Quoted in Lanouette, *Genius in the Shadows*, 308; and Bernstein, "Dazzling Gadfly," 27.

16. Quoted in Brodie and Brodie, *From Crossbow to H-Bomb*, 253.

17. See Walsh, *Scientists as Prophets*, 109 and 113; and Wang, *American Science in an Age of Anxiety*, 44, 45, and 59.

18. Although a dramatization, see Michael Frayn's characterizations of atomic physicists Niels Bohr and Werner Heisenberg for a thoughtful representation of atomic scientists' confrontation with Techne's Paradox (*Copenhagen*, 65–66).

19. Szilard, *His Version of the Facts*, xvii.

20. On Szilard's conflicted motivations, see Spencer R. Weart and Gertrude Weiss Szilard's preface to *His Version of the Facts* (xvii).

21. Szilard, *His Version of the Facts*, vi. The ten commandments "reflect Szilard's spirit like a portrait," according to his editors (xviii).

22. Aristotle, *On Rhetoric*, 38 [1.1.4].

23. Godin, "Writing Performative History," 469.

24. The norms of scientific ethos were named by sociologist of science Robert Merton and later developed as rhetorical *topoi* by Lawrence Prelli. See Merton, *Sociology of Science*, 258–59 and 268–70; and Prelli, "Rhetorical Construction of Scientific Ethos."

25. Aristotle, *On Rhetoric*, 38 [1.1.4].

26. Prelli, "Rhetorical Construction of Scientific Ethos," 50.

27. According to Szilard's friend Edward Shils, Szilard spoke of "everything in a disinterested and precise way" ("Leo Szilard: A Memoir," 37).

28. Szilard, *His Version of the Facts*, 46.

29. Szilard to Enrico Fermi, March 13, 1936, Leo Szilard Papers.

30. Szilard, *His Version of the Facts*, 47.

31. Szilard, *His Version of the Facts*, 47.

32. Szilard, *His Version of the Facts*, 41–42.

33. Szilard, *His Version of the Facts*, 23. The Bund represented the first of many such endeavors, including his much-later, and more-successful, work to help organize the "Pugwash Movement," "The Emergency Committee of Atomic Scientists," the "Council for a Livable World," and other abortive endeavors like the "Angels Project," which Szilard proposed in 1962 as a way to bring together US and Soviet policy-makers to stop the arms race (22).

34. Szilard to Vannevar Bush, July 8, 1942, Leo Szilard Papers.

35. Szilard, *His Version of the Facts*, 164.

36. Szilard, *His Version of the Facts*, 164.

37. Szilard, *His Version of the Facts*, 176.

38. Szilard, *His Version of the Facts*, 191–92, emphasis in original. Szilard's editors noted that Szilard might not have authored this document but believed he probably did (189).

39. After Roosevelt's death, Szilard revised the memo for his meeting with James Byrnes, President Truman's soon-to-be secretary of state.

40. Szilard, *His Version of the Facts*, 206.

41. Szilard, *His Version of the Facts*, 206.

42. Szilard, *His Version of the Facts*, 206.

43. Szilard, *His Version of the Facts*, 187.

44. Szilard, *His Version of the Facts*, 210.

45. Szilard, *His Version of the Facts*, 210.

46. Szilard, *His Version of the Facts*, 210.

47. "More than 700" scientists worked at the Los Alamos wing of the Manhattan Project alone during 1945 (V. Jones, *Manhattan*, 348).

48. Szilard, *His Version of the Facts*, 213 and 224.

49. Merton, *Sociology of Science*, 261.

50. A. Smith, "Elusive Dr. Szilard," 77.

51. On Szilard's differentiation between the truthful realm of scientific communication and the untruthful realm of political communication, see Bess, *Realism, Utopia, and the Mushroom Cloud*, 69; Lanouette, *Genius in the Shadows*, 450; Szilard, *His Version of the Facts*, vi; Szilard, *Scientific Papers*, xix; Szilard, *Voice of the Dolphins*, 53–54; "Address by Dr. Leo Szilard," press release from the *Nation*, December 3, 1945, Leo Szilard Papers; and Szilard, letter to "Colleague," October 24, 1956, Leo Szilard Papers. He later revised this letter for inclusion in his short story "The Voice of the Dolphins."

52. On the control of scientists by the military, and the exclusion of scientists from atomic policy-making, see Hoch's "Crystallization of a Strategic Alliance" and Smith's *Peril and a Hope*.

53. Szilard, *His Version of the Facts*, 136. In May 30, 1940, correspondence with physicists Louis A. Turner and Harold C. Urey, Szilard proposed that "free discussion of all results and ideas among as many physicists as is practicable should not be inhibited," although all participants must be "trustworthy" and pledge to withhold information from everyone else (128 and 130). Turner, among others—including his frequent partner Fermi and French physicist Frédéric Joliot—did not comply with Szilard's plan. Szilard expressed incredulity about his colleagues' claims that nuclear fission had already gained international fame when he was the one responsible for disseminating the "dangerous manuscripts" throughout Europe and the United States (57, 69–70, and 77).

54. Szilard, *His Version of the Facts*, 118.

55. Szilard, *His Version of the Facts*, 115.

56. Szilard, *His Version of the Facts*, 121.

57. Szilard, *His Version of the Facts*, 141.

58. Szilard, *His Version of the Facts*, 140.

59. Szilard, *His Version of the Facts*, 141.

60. Hoddeson, et al., *Critical Assembly*, 95; Rhodes, *Making of the Atomic Bomb*, 504.

61. Szilard to A. H. Compton, December 3, 1942, Leo Szilard Papers.

62. Szilard, *His Version of the Facts*, 163.

63. Quoted in Sherwin, *World Destroyed*, 117.

64. Quoted in Rhodes, *Making of the Atomic Bomb*, 502; and Sherwin, *World Destroyed*, 116–17.

65. According to Gertrude Weiss Szilard, Szilard's wife and literary executor, he "had a sense of history . . . and carefully preserved all notebooks, correspondence, and drafts of his writings and even filed documents of special importance in folders marked 'History'" (Szilard, *Scientific Papers*, xix).

66. Szilard, *Toward a Livable World*, 207.

67. As much as science is a constructed representation of nature, scientific "facts and fictions coexist as siblings" (Peters and Rothenbuhler, "Role of Rhetoric in the Making of a Science," 17).

68. Szilard, *His Version of the Facts*, 149, emphasis in original.

69. Szilard, *His Version of the Facts*, 3.

70. Perelman and Olbrechts-Tyteca, *New Rhetoric*, 116.

71. Szilard, *Voice of the Dolphins*, 106.

72. Shapin and Schaffer, *Leviathan and the Air-Pump*.

73. Kuhn, *Structure of Scientific Revolutions*; Hacking, *Social Construction of What?* 80–92.

74. MacKenzie, *Inventing Accuracy*, 381 and 418; Latour, *Science in Action*, 100. On the use of facts and truths in argumentation and rhetoric, see Perelman and Olbrechts-Tyteca, *New Rhetoric*, 67–70.

75. Jamison and Eyerman, *Seeds of the Sixties*, 105, 113, 116–18, and 138. See also Gruber, "Manhattan Project Maverick," 73.

76. A. Smith, *Peril and a Hope*, 359–60.

77. For a history of twentieth-century deterrence theory leading up to the end of World War II, see Quester's *Deterrence before Hiroshima*.

78. Robert L. Scott argued that the oxymoronic character of the term "Cold War" imbued the concept of the Cold War with ambivalence such that the term came to mean fighting a war by "stopping short" in both "words and actions," which thereby kept the nuclear powers in check ("Cold War and Rhetoric," 4). See also Szilard, *Toward a Livable World*, 409.

79. Kahn, "On the Nature and Feasibility of War and Deterrence," 28, emphasis added.

80. Wohlstetter, "Delicate Balance of Terror," emphasis added.

81. Wohlstetter, "Delicate Balance of Terror," 212.

82. Kavka, *Moral Paradoxes of Nuclear Deterrence*, 15.

83. Lawrence Freedman wrote, "Deterrence can be a technique, a doctrine and a state of mind" (*Deterrence*, 116).

84. For a history of these tests, see Miller's *Under the Cloud*.

85. Bryan Taylor, "Bodies of August," 332.

86. Derrida, "No Apocalypse, Not Now," 23.

87. Szilard, *Toward a Livable World*, 238.

88. Kahn, *On Thermonuclear War*, 9.

89. Szilard, *His Version of the Facts*, 152.

90. Szilard, *His Version of the Facts*, 154. Szilard hereby expressed a war cliché.

91. Laurence, "Drama of the Atomic Bomb," 1.

92. Szilard, *His Version of the Facts*, 163.

93. Sherwin, *World Destroyed*, 288.

94. Szilard, *His Version of the Facts*, 163.

95. Szilard, *Toward a Livable World*, 132.

96. Szilard, *His Version of the Facts*, 53.

97. Fisher, "Narration as a Human Communication Paradigm," 13.

98. Historian Guy Oakes used the term "imaginary war" to refer to the US government's civil defense programs that sought to normalize the prospect of nuclear war for Americans (*Imaginary War*, 33). See also Mechling and Mechling, "Campaign for Civil Defense."

99. Chernus, *Apocalypse Management*, 2.

100. Fisher, "Narration as a Human Communication Paradigm," 13. See also Goodnight, "Personal, Technical, and Public Spheres of Argument."

101. Szilard, *His Version of the Facts*, xvii.

102. Szilard, *His Version of the Facts*, 37.

103. Szilard, *His Version of the Facts*, 154.

104. Szilard, *His Version of the Facts*, 197–98, and 211.

105. Fisher, "Narration as a Human Communication Paradigm," 12–13.

106. Perelman and Olbrechts-Tyteca, *New Rhetoric*, 67.

107. Hikins, "Rhetoric of 'Unconditional Surrender.'"

108. Szilard, *Toward a Livable World*, 407 and 417. Szilard first proposed minimal deterrence at the first Pugwash Conference in 1957 (Lanouette, *Genius in the Shadows*, 371). Minimal deterrence is also known as minimum deterrence.

109. On "massive retaliation" as a "vague and blunt" rhetorical strategy that kept the international community guessing about the United States' nuclear intentions, see Schaefermeyer's "Dulles and Eisenhower on 'Massive Retaliation'" (28).

110. O'Gorman, *Spirits of the Cold War*, 76–79.

111. O'Gorman, *Spirits of the Cold War*, xvii.

112. Szilard, *Toward a Livable World*, 215.

113. Szilard, *Voice of the Dolphins*, 162.

114. Hilgartner, Bell, and O'Connor, *Nukespeak*, xiii. See also Schiappa's "Rhetoric of Nukespeak."

115. See Szilard's "'Minimal Deterrent' vs. Saturation Parity" in *Toward a Livable World* (407–21).

116. Szilard, *Toward a Livable World*, 219.

117. Szilard, *Toward a Livable World*, 407, emphasis in original.

118. Szilard, *Toward a Livable World*, 421.

119. Bernstein, "Introduction" in Szilard, *Voice of the Dolphins*, 15.

120. Szilard, *Toward a Livable World*, 418.

121. Lanouette, *Genius in the Shadows*, 444; Lewis "From Science to Science Fiction," 99.

122. Lewis, "From Science to Science Fiction," 96.

123. Szilard, *Voice of the Dolphins*, 160.

124. Perelman and Olbrechts-Tyteca, *New Rhetoric*, 118.

125. Disch, *Dreams Our Stuff Is Made Of*, 12–13.

126. Canaday, *Nuclear Muse*, 228, 230, and 233.

127. Szilard, *Toward a Livable World*, 230.

128. Bryan Taylor, "Home Zero," 212.

129. In *Ambient Rhetoric*, Rickert wrote that the Heideggerian concept of *dwelling* "places us in the insight that rhetoric, being worldly, cannot be understood solely as human doing and that persuasion gains its bearings from an affectability that emerges with our material environments both prior to and alongside the human" (254).

130. See Farrell and Goodnight, "Accidental Rhetoric"; Goodnight, "On Questions of Evacuation and Survival in Nuclear Conflict"; Hogan, *Nuclear Freeze Campaign*; Ivie, "Metaphor"; and King and Petress, "Universal Public Argument and the Failure of the Nuclear Freeze."

131. Szilard never wrote of pragmatic idealism, but his friend and colleague Jonas Salk identified its presence in Szilard's work. Salk reflected, "In his special ways, through a quest for knowledge and through the force of his pragmatic idealism, he sought to create a more peaceful world" (Lanouette, *Genius in the Shadows*, xiii). Additionally, physicist and Szilard collaborator Bernard Feld wrote that "Szilard's 'pacifism' was always less idealistic and more pragmatic than that of his mentor and collaborator," Einstein ("Leo Szilard, Scientist for All Seasons," 676).

132. Aristotle synthesized the squabbling of Plato and Gorgias over the definition of rhetoric by defining it as the counterpart of philosophical dialectic, which brought pragmatics and ideals into the same realm of activity (*On Rhetoric*, 28 [1354a]). Furthermore, Aristotle defined *phronēsis* as the action of a prudent individual who identifies the available means of persuasion as an act of creating a better world based on a lofty ideal (*On Rhetoric*, 36 [1355a]). A prudent individual must balance the reasonable with the desirable, and the desirable with the practical (Self, "Rhetoric and Phronesis," 134). Rescher named and began

elucidating pragmatic idealism as a philosophical school in *A System of Pragmatic Idealism*. To describe how the usually differentiated concepts of pragmatism and idealism mingle, Rescher wrote that "insofar as realism stands on [a] pragmatic basis, it does not rest on considerations of independent substantiating evidence about how things actually stand in the world, but rather it is established by considering, as a matter of practical reasoning, how we do (and must) think about the world within the context of the projects to which we stand committed" ("Pragmatic Idealism and Metaphysical Realism," 397). Rescher indicated that these discrete epistemological and political concepts overlap by informing each other's development in practice.

133. Collins and Pinch, *Golem at Large*, 155.

134. See O'Gorman, "Eisenhower and the American Sublime," 46.

135. O'Gorman, "Eisenhower and the American Sublime," 56.

136. Medhurst, "Eisenhower's 'Atoms for Peace,'" 205, 209, and 213–18.

137. A. Smith, *Peril and a Hope*, 160.

138. Aron, *Century of Total War*, 149.

Chapter 5

The epigraphs to this chapter are drawn from Kaczynski, *Communiques of Freedom Club*, 16, and Wright, "Evolution of Despair."

1. My biographical portrayal of Kaczynski is based on Alston Chase's *A Mind for Murder*.

2. Gelernter wrote about the attack in *Drawing Life: Surviving the Unabomber*.

3. Kaczynski, *Communiques of Freedom Club*, 16.

4. Kaczynski, *Communiques of Freedom Club*, 16.

5. Kaczynski, *Technological Slavery*, 38.

6. Kaczynski, *Technological Slavery*, 13.

7. Kaczynski, *Technological Slavery*, 104.

8. Kaczynski, *Technological Slavery*, 69 and 77.

9. So, too, does the rhetoric of Earth First!—a group that Kaczynski followed and with which he is sometimes associated—"refuse to compromise" (Lange, "Refusal to Compromise"). On the somewhat tenuous link between Kaczynski and Earth First!, see Taylor's "Religion, Violence and Radical Environmentalism" (28–30).

10. Kaczynski's bombing campaign took place around the time when the neologism "improvised explosive device" was coined and became more commonplace (Gill, Horgan, and Lovelace, "Improvised Explosive Device," 733–34). I use "IED" to describe his bombs to encompass the different types of bombs he made, and to fit with the current understanding of what IEDs are and how to define them. See also The National Academies and The Department of Homeland Security's *IED Attack*.

11. Perelman and Olbrechts-Tyteca, *New Rhetoric*, 91.

12. Kaczynski, *Technological Slavery*, 375.

13. Kaczynski, *Technological Slavery*, 38, 91, and 97, emphasis in original. See also his essay "The Truth about Primitive Life: A Critique of Anarcho-Primitivism," in which Kaczynski lauds the primitive life but clarifies that primitive life is far from utopic (Kaczynski, *Technological Slavery*, 126–89).

14. See Skrbina, "Revolutionary for Our Times," 27.

15. The abstract for *Boundary Functions* and a bibliography of eight mathematics papers published by Kaczynski are available at Bullough, "Published Works of Theodore Kaczynski."

16. Kaczynski, *Communiques of Freedom Club*, 3. For the technical design of his pipe bombs, see his letter to the *San Francisco Examiner* (Kaczynski, *Communiques of Freedom Club*, 4).

17. Kaczynski, *Technological Slavery*, 65.

18. Kaczynski, *Communiques of Freedom Club*, 8.

19. Kaczynski, "Hit Where It Hurts"; Kaczynski, *Technological Slavery*, 227.

20. Kaczynski, *Communiques of Freedom Club*, 7.

21. Kaczynski, *Communiques of Freedom Club*, 10. In addition to *Industrial Society and Its Future*, Kaczynski tried a number of different rhetorical approaches and genres to spread his message to the American reading public. He penned a short essay, "The Wave of the Future," for the *Saturday Review* that, dripping with irony, touted the benefits of speculative inventions such as the manipulation of clouds to create mass-mediated entertainment in the sky. He penned satirical fiction with the short story "Ship of Fools." He penned his famous manifesto and a number of other essays. He penned a lot of letters.

22. Regarding the portrayal of Kaczynski's mental status, Tim Luke noted that, to say the least, "analyses indulging in psycho-babble [are] not lacking" ("Re-Reading the Unabomber Manifesto," 82).

23. J. Johnson, "Skeleton on the Couch," 463. As rhetorician Catherine Prendergast put it, "the diagnosis of schizophrenia necessarily supplants one's position as a rhetor" so that "to be disabled mentally is to be disabled rhetorically" ("On the Rhetorics of Mental Disability," 47 and 57).

24. Pryal, "Genre of Mood Memoir and the *Ethos* of Psychiatric Disability," 480; Prendergast, "On the Rhetorics of Mental Disability," 57. See also Molloy's "Recuperative Ethos and Agile Epistemologies."

25. Johnson, "Skeleton on the Couch," 475.

26. Prendergast, "On the Rhetorics of Mental Disability," 56. Prendergast evidenced Kaczynski as someone who was denied rhetoricability because of his schizophrenia diagnosis (56–57).

27. Perelman and Olbrechts-Tyteca, *New Rhetoric*, 91. See also Perelman's *New Rhetoric and the Humanities* (160–61).

28. Perelman and Olbrechts-Tyteca, *New Rhetoric*, 92.

29. Perelman and Olbrechts-Tyteca, *New Rhetoric*, 92.

30. Perelman and Olbrechts-Tyteca, *New Rhetoric*, 92.

31. Kaczynski, *Technological Slavery*, 78.

32. Kaczynski, "Answer to Some Comments," 2.

33. Perelman, *New Rhetoric and the Humanities*, 160.

34. Kaczynski expressed dismay that critics of *Industrial Society and Its Future* spent so much time disapproving of his writing style rather than addressing his arguments about technology (Kaczynski, *Technological Slavery*, 124).

35. A more contemporary example of soft machine rhetoric is Ray Kurzweil's ongoing technofuturist project that seeks the integration of human biology and technology ("Promise and Peril," 3).

36. Kaczynski, *Technological Slavery*, 14–15, 108, 111, and 253; Kaczynski, "Ship of Fools," 453–54.

37. Kaczynski, *Technological Slavery*, 39, 42, and 110. Kaczynski's critique of tolerance closely resembles that of Herbert Marcuse's "Repressive Tolerance."

38. Kaczynski, "Ship of Fools," 453–54; Kaczynski, *Technological Slavery*, 199.

39. Kaczynski, "Answer to Some Comments," 3.

40. Kaczynski, *Technological Slavery*, 14 and 229–30.
41. Kaczynski, "Ship of Fools," 456–57.
42. Kaczynski, *Technological Slavery*, 362.
43. Zerzan, *Twilight of the Machines*, 97.
44. Perelman and Olbrechts-Tyteca, *New Rhetoric*, 92.
45. Perelman and Olbrechts-Tyteca, *New Rhetoric*, 92.
46. Luke, "Re-Reading the Unabomber Manifesto," 92.
47. Kaczynski, *Technological Slavery*, 40.
48. Kaczynski, *Technological Slavery*, 251.
49. Kaczynski, *Technological Slavery*, 251.
50. Wolff, "Beyond Tolerance," 4.
51. Wolff, "Beyond Tolerance," 52.
52. Wolff, "Beyond Tolerance," 52.
53. Marcuse, "Repressive Tolerance," 83; see also 93–95, 107, and 122–23.
54. Clausewitz, *On War*, 77.
55. Kaczynski, *Technological Slavery*, 268.
56. De Kerckhove, "On Nuclear Communication," 80.
57. See Marcuse, "Repressive Tolerance," 85.
58. Kaczynski, *Technological Slavery*, 251.
59. Kaczynski, *Communiques of Freedom Club*, 8; Kaczynski, *Technological Slavery*, 38–39 and 125.
60. Skrbina, "Technological Anarchism."
61. In his account of the bombing, Gelernter claimed that he is anti-computers and anti-technology by raising the possibility of some computer-based problems in the epilogue to *Mirror Worlds*. Yet his promotion of computer software, despite its possible negative effects on humanity, displays his technological ambivalence (*Drawing Life*, 28, 37, and 59). Owing to his questioning of technology's disadvantages, Gelernter asserted that Kaczynski's letter indicated that the bomber had chosen the wrong target (27–28), although Kaczynski's letter to Gelernter reaffirmed his distaste for Gelernter's work. Anarchist John Zerzan confirmed the reasons for Kaczynski's attack on Gelernter in *Running on Emptiness* (154–55).
62. Gelernter, *Mirror Worlds*, 3, emphasis in original.
63. Gelernter, *Mirror Worlds*, 216 and 224. Gelernter's ambivalent soft rhetoric is prevalent in the book's epilogue.
64. Kaczynski, "Answer to Some Comments," 2.
65. Cox, "Die Is Cast," 230.
66. Kaczynski, *Technological Slavery*, 76–77.
67. Kaczynski, "Unnamed Essay."
68. Kaczynski, *Technological Slavery*, 117.
69. Kaczynski, *Technological Slavery*, 117.
70. Kaczynski, *Technological Slavery*, 291.
71. Kaczynski, *Technological Slavery*, 291.
72. Kaczynski, *Technological Slavery*, 291.
73. Kaczynski, "Why the Technological System Will Destroy Itself," 4 and 13.
74. Kaczynski, *Technological Slavery*, 320.
75. Kaczynski, *Technological Slavery*, 320.
76. Ellul, "Technological Order," 102–3.
77. Kaczynski, *Technological Slavery*, 74–75 (see also 278, 286, and 315). See also Kaczynski's letter to *Scientific American* (*The Communiques of Freedom Club*, 11–12).

78. Kaczynski, *Technological Slavery*, 74.
79. Kaczynski, *Technological Slavery*, 75.
80. Kaczynski, *Technological Slavery*, 300.
81. Kaczynski, *Technological Slavery*, 303.
82. Kaczynski, *Technological Slavery*, 62–64.
83. Kaczynski, *Technological Slavery*, 62–63.
84. Kaczynski, *Technological Slavery*, 63.
85. Kaczynski, *Technological Slavery*, 63.
86. Kaczynski, *Technological Slavery*, 63.
87. Kaczynski, *Technological Slavery*, 63.
88. Kaczynski, *Technological Slavery*, 63.
89. Chase, *Mind for Murder*, 32 and 293.
90. Chase, *Mind for Murder*, 204–5 and 209–13. Skrbina's "Technological Anarchism" situates Kaczynski's thought within twentieth-century philosophy of technology, and Corey's "On the Unabomber" compares and contrasts Ellul and Kaczynski.
91. Kaczynski, *Technological Slavery*, 124.
92. See Chase, *Mind for Murder*, 228–94.
93. Kaczynski, *Technological Slavery*, 212.
94. Kaczynski, *Technological Slavery*, 212.
95. Kaczynski, *Technological Slavery*, 98.
96. Kaczynski, *Technological Slavery*, 291.
97. Kaczynski, "Unnamed Essay."
98. Kaczynski, *Technological Slavery*, 77.
99. Cox, "Die Is Cast," 227.
100. Kaczynski, like the poet William Wordsworth, wanted the "restoration of our humanity" (Lentricchia and McAuliffe, *Crimes of Art + Terror*, 19–20).
101. Kaczynski, *Technological Slavery*, 13.
102. Kaczynski, *Technological Slavery*, 13.
103. Kaczynski, *Technological Slavery*, 317.
104. Kaczynski, *Technological Slavery*, 317.
105. Kaczynski, *Technological Slavery*, 13.
106. Kaczynski, *Technological Slavery*, 14.
107. Kaczynski, *Technological Slavery*, 66.
108. Kaczynski, *Technological Slavery*, 38.
109. Kaczynski, *Technological Slavery*, 14.
110. Cox, "Die Is Cast," 234 and 236.
111. Zarefsky, *Rhetorical Perspectives on Argumentation*, 95.
112. Kaczynski, *Technological Slavery*, 13 and 96.
113. Kaczynski, *Technological Slavery*, 216.
114. Kaczynski, *Technological Slavery*, 90.
115. Cox, "Die Is Cast," 229.
116. Skrbina, "Technological Anarchism," 104.
117. Chase, *Mind for Murder*, 100.
118. See Kaczynski, *Communiques of Freedom Club*, 8.
119. Lentricchia and McAuliffe, *Crimes of Art + Terror*, 19 and 23.
120. On mere "men of words," see Kaczynski, *Technological Slavery* (355–56); and Kaczynski "Hit Where It Hurts" (8).
121. Baudrillard, *Spirit of Terrorism*, 13.

122. The FBI coined the term "UNABOM," sometimes fully capitalized, with reference to the unknown bomber's crimes: "'Un' for universities, 'a' for airlines, and 'bom' for bomb" (Chase, *Mind for Murder*, 55).

123. Pickering, *Mangle of Practice*, 24.

124. This sentence paraphrases John Tresch's description in *Romantic Machine* of the protagonist in a Balzac novel, *Lost Illusions*, whose experience with the press leads to ruin, and although the example comes from far afield, it is an apt description of Kaczynski's fate after his violent mobilization of newsprint (xv–xvi).

125. Butler, *Excitable Speech*, 4. See also Scarry's *Body in Pain* (43).

126. Yellin, *Battle Exhortation*.

127. Scarry, *Body in Pain*, 67.

128. Bolt, *Violent Image*, 199–224.

129. See the nonviolent "revolutionary plan" against "technological society" that Ellul described in *Autopsy of Revolution* (281–91).

130. Kaczynski, "Hit Where It Hurts," 6 and 8.

131. Kaczynski, "Hit Where It Hurts," 3.

132. Kaczynski, "Hit Where It Hurts," 3.

133. Kaczynski, "Hit Where It Hurts," 4.

134. Kaczynski, *Technological Slavery*, 251 and 253, emphasis in original.

135. Kaczynski, "Hit Where It Hurts," 8.

136. Kaczynski, "Hit Where It Hurts," 8.

137. Galison, "Image of Self," 257.

138. Galison, "Image of Self," 259.

139. Historian Tami Davis Biddle made this argument regarding the invention of the airplane (*Rhetoric and Reality in Air Warfare*, 289).

140. Amador and Paul-Odouard, "Defending the Unabomber," 368 and 364. See also S. Johnson, "Psychiatric Competency Report," 24; Kaczynski's assessment of his legal treatment in "Explanations of the Judicial Opinions" (Kaczynski, *Technological Slavery*, 410–14); and Kirk and Kutchins, *Selling of DSM*, 20–22.

141. Kaczysnki's psychological profiling exemplifies rhetorician Cathryn Molloy's argument that one type of psychiatric "over-diagnosis" involves "the language and rhetoric of esoteric medical expertise eclipsing the voices and experiences of mentally ill persons" ("Recuperative Ethos and Agile Epistemologies," 140).

142. Mello, "Non-Trial of the Century," 505–7.

143. Skrbina, "A Revolutionary for Our Times," 34.

144. For a contemporaneous critique of the American Psychiatric Association's *Diagnostic and Statistical Manual of Mental Disorders* (*DSM*), see Kutchins and Kirk's *Making Us Crazy* (21–54).

145. Daniel Kevles wrote, "The Unabomber combines the views of the Luddites with the cruelty of their repressors; to him, industrialism is not only an offense, it is an offense punishable by death" ("E Pluribus Unabomber," 2).

146. Material-culture scholar and archaeologist Ian Hodder defined "entanglement" as the process in which "the social world of humans and the material world of things are entangled together by dependences and dependencies that create potentials, further investments and entrapments" (*Entangled*, 89 and 95).

147. Benson, "Unabomber and the History of Science"; Shrum, "We Were the Unabomber"; Restivo, "4S, the FBI, and Anarchy."

148. Vollmann, "Machines of Loving Grace," 69–70.

149. On the ideological and verbal links between Kaczynski and these environmental movements, see Taylor's "Religion, Violence and Radical Environmentalism." In *The Pyrotechnic Insanitarium,* Mark Dery argued that Kaczynski's appeals to wild nature have more in common with the social-Darwinistic appeals of the probusiness elite who mobilize nature to justify their nonhumanitarian goals (227–45).

150. S. Jones, *Against Technology,* 212.

151. Kaczynski, "Hit Where It Hurts," 3.

152. For an assessment of Kaczynski by comparison to Luddite and neo-Luddite thinking, see Steven E. Jones's *Against Technology* (211–33).

153. Kaczynski, letter to the *San Francisco Chronicle,* reprinted in M. Taylor, "S.F., L.A. Airports Get Bomb Warning."

154. Derrer, *We Are All the Target,* 5.

155. Packer, "Becoming Bombs."

156. Sloterdijk, *Terror from the Air,* 14 and 16.

157. See Rozelle's *Ecosublime* (88).

158. Joy, "Why the Future Doesn't Need Us."

159. Oleson, "'Evil the Natural Way.'" Oleson suggested that Thoreau might have approved of Kaczynski's bombings depending on whether the mad bomber or the philosopher-mathematician was deemed responsible (224–25).

160. Quoted in Chase, *Mind for Murder,* 32.

161. Chase, *Mind for Murder,* 128.

162. Baudrillard, *Spirit of Terrorism,* 5.

163. K. Kelly, *What Technology Wants,* 199.

164. Benson, "Unabomber and the History of Science," 104–5.

165. Kaczynski's social-Darwinistic tendencies are evident in "Why the Technological System Will Destroy Itself," an essay in which he conceived of a type of technological Darwinism.

166. Kaczynski, "When Non-Violence Is Suicide," 2.

167. On the rhetorical self-portraiture of weapons advocates, see Medhurst's "Rhetorical Portraiture" (52).

168. Restivo, "4S, the FBI, and Anarchy," 90.

Conclusion

The first epigraph for this chapter is reproduced in Kelly, *Manhattan Project,* 370; the second is drawn from Plato, *Socrates' Defense (Apology),* 8–9 [22d–e].

1. McCullough, "Higgs Adventure."

2. Future of Life Institute, "Autonomous Weapons."

3. Turkle, *Alone Together;* Schüll, *Addiction by Design.*

4. On the rhetorical plasticity of amplification, see Ian Hill's "Not Quite Bleeding from the Ears."

5. See Wander, "Rhetoric of American Foreign Policy," 174.

6. Arendt, "Rand School Lecture," 221.

7. On the "rhetoric of progress," see Keith and Zagacki's "Rhetoric and Paradox in Scientific Revolutions" (175). Kaczynski's essay "The Truth about Primitive Life" begins with a stinging critique of the "myth of progress" (Kaczynski, *Technological Slavery,* 128). Technology scholar Carl Mitcham asserted that "the history of technology is not nearly so linear and

progressive as technological history implies" (*Thinking through Technology*, 134). On the rhetorical appeal to seek technological solutions to technological problems with respect to weapons systems, see Holloway's "Strategic Defense Initiative" (210–11 and 228). On the "sterilization" of weapons to purge them of their dyslogistic connotations, see Ott, Aoki, and Dickinson's "Ways of (Not) Seeing Guns" (223).

8. Paul Virilio argued throughout his oeuvre that weaponry and language globalize violence by conflating speed, time, space, and information.

9. See M. Benjamin, *Drone Warfare*, 200; Hasian, *Drone Warfare and Lawfare in a Post-Heroic Age*; Packer and Reeves, "Romancing the Drone," 309–10; US Department of Defense, the Minerva Initiative; US Department of Justice, "Procedures for Approving Direct Action against Terrorist Targets," 3.D.3.3, 1.G.2 and 4.A–C, and 5.A–5.B; and Walters, "Drone Strikes, *Dingpolitik* and Beyond," 113–14.

10. Bjork, *Strategic Defense Initiative*, 1; Benson and Anderson, "Ultimate Technology," 257.

11. Atwill, *Rhetoric Reclaimed*, 48 and 96.

12. Aristotle, *On Rhetoric*, 36 [1355b].

13. Plato, *Socrates' Defense (Apology)*, 8–9 [22d–e].

14. Reproduced in C. Kelly, *Manhattan Project*, 370.

15. Jünger wrote that "the emotions of the heart and the systems of the intellect can be disproved, but an object cannot be disproved—and such an argument is a machine gun" (quoted in Rohkrämer, "How Ernst Jünger Learned to Love Total War," 181).

16. See Perelman and Olbrechts-Tyteca, *New Rhetoric*, 117.

17. See Eden, et al., "Three Minutes and Counting."

18. Canetti, *Crowds and Power*, 25.

Bibliography

Aberth, John. *From the Brink of the Apocalypse: Confronting Famine, War, Plague, and Death in the Later Middle Ages*. New York: Routledge, 2002.

Adamic, Louis. *Dynamite: The Story of Class Violence in America*. Rev. ed. New York: Viking Press, 1934.

Agamben, Giorgio. *Homo Sacer: Sovereign Power and Bare Life*. Translated by Daniel Heller-Roazen. Stanford: Stanford University Press, 1998.

Akrich, Madeline. "The De-Scription of Technical Objects." In *Shaping Technology/Building Society: Studies in Sociotechnical Change*, edited by Wiebe E. Bijker and John Law, 205–24. Cambridge, Mass: MIT Press, 1992.

Allen, Hervey. *Toward the Flame: A Memoir of World War I*. Lincoln: University of Nebraska Press, 2003.

Allen, Robert S. "Chemical Warfare, a New Industry." *The Nation*, January 12, 1927, 33.

Altgeld, John P. "Altgeld's Reasons for Pardoning Fielden, Neebe and Schwab." In *The Chicago Martyrs: The Famous Speeches of the Eight Anarchists in Judge Gary's Court, October 7, 8, 9, 1886, and Reasons for Pardoning Fielden, Neebe and Schwab*, 131–59. San Francisco: Free Society, 1899.

Amador, Xavier F., and Reshmi Paul-Odouard. "Defending the Unabomber: Anosognosia in Schizophrenia." *Psychiatry Quarterly* 71, no. 4 (2000): 363–71.

Arendt, Hannah. "Rand School Lecture." In *Essays in Understanding, 1930–1954: Formation, Exile, and Totalitarianism*, 217–27. New York: Schocken, 1994.

Aristotle. *Metaphysica*. Translated by W. D. Ross. In *The Basic Works of Aristotle*, edited by Richard McKeon, 681–926. New York: Modern Library, 2001.

———. *On Rhetoric: A Theory of Civic Discourse*. Translated by George A. Kennedy. New York: Oxford University Press, 1991.

Aron, Raymond. *The Century of Total War*. Garden City: Doubleday, 1954.

Atwill, Janet. *Rhetoric Reclaimed: Aristotle and the Liberal Arts Tradition*. Ithaca: Cornell University Press, 1998.

Auld, S. J. M. *Gas and Flame in Modern Warfare*. New York: George H. Doran, 1918.

"The Aurora Turnverein." In *German Workers in Chicago: A Documentary History of Working-Class Culture from 1850 to World War I*, edited by Hartmut Keil and John B. Jentz, 160–69. Urbana: University of Illinois Press, 1988.

Avery, John. *Progress, Poverty and Population: Re-Reading Condorcet, Godwin and Malthus*. Portland: Frank Cass, 1997.

Avrich, Paul. *The Haymarket Tragedy*. Princeton: Princeton University Press, 1984.

Bacon, Francis. *Advancement of Learning*. Edited by J. Devey. New York: American Home Library, 1902.

Bakunin, Michael. *God and the State*. New York: Dover, 1970.

———. *Statism and Anarchy*. Translated and edited by Marshall Shatz. New York: Cambridge University Press, 1990.

Barad, Karen. *Meeting the Universe Halfway: Quantum Physics and the Entanglement of Matter and Meaning*. Durham: Duke University Press, 2006.

Barnard, Harry. *Eagle Forgotten: The Life of John Peter Altgeld*. New York: Duell, Sloan and Pearce, 1938.

Barnett, Scot. "Chiasms: Pathos, Phenomenology, and Object-Oriented Rhetorics." *Enculturation*, no. 20. Last updated November 23, 2015. http://enculturation.net/chiasms-pathos-phenomenology.

Barthes, Roland. *The Semiotic Challenge*. Translated by Richard Howard. Berkeley: University of California Press, 1994.

Battlestar Galactica. SyFy Channel, 2004–9.

Baudrillard, Jean. *The Spirit of Terrorism and Other Essays*. Translated by Chris Turner. New York: Verso, 2003.

Bazerman, Charles. "The Production of Technology and the Production of Meaning." *Business and Technical Communication* 12, no. 3 (1998): 381–87.

Beales, H. L. "The Historical Context of the *Essay* on Population." In *Introduction to Malthus*, edited by D. V. Glass, 3–24. New York: Wiley & Sons, 1953.

Beck, Ulrich. *Ecological Enlightenment: Essays on the Politics of the Risk Society*. Atlantic Highlands: Humanities Press, 1995.

———. *World Risk Society*. Malden: Polity, 1999.

Bekken, Jon. "The First Anarchist Daily Newspaper: The *Chicagoer Arbeiter-Zeitung*." *Anarchist Studies* 3, no. 1 (1995): 3–23.

Bell, David A. *The First Total War: Napoleon's Europe and the Birth of Warfare as We Know It*. New York: Houghton Mifflin, 2007.

Benjamin, Medea. *Drone Warfare: Killing by Remote Control*. New York: Verso, 2013.

Benjamin, Walter. "Theses on the Philosophy of History." In *Illuminations: Essays and Reflections*, 253–64. Translated by Harry Zohn. New York: Schocken, 1968.

Bennett, Jane. "Powers of the Hoard: Artistry and Agency in a World of Vibrant Matter." Recorded on September 13, 2011, in New York. YouTube video, 1:14:44. https://www.youtube.com/watch?v=q607Ni23QjA.

———. *Vibrant Matter: A Political Ideology of Things*. Durham: Duke University Press, 2010.

Bennett, Tony, and Patrick Joyce, eds. *Material Powers: Cultural Studies, History and the Material Turn*. New York: Routledge, 2010.

Benson, Keith. "The Unabomber and the History of Science." *Science, Technology, and Human Values* 26, no. 1 (2001): 101–5.

Benson, Thomas W., and Carolyn Anderson. "The Ultimate Technology: Frederick Wiseman's *Missile*." In *Communication and the Culture of Technology*, edited by Martin J. Medhurst, Alberto Gonzalez, and Tarla Rai Peterson, 257–83. Pullman: Washington State University Press, 1990.

Bernstein, Barton J. "Dazzling Gadfly." *Inquiry*, January 8 and 22, 1979.

———. "Introduction." In *The Voice of the Dolphins and Other Stories*, by Leo Szilard, expanded ed., 1–43. Stanford: Stanford University Press, 1992.

Bess, Michael. *Realism, Utopia, and the Mushroom Cloud: Four Activist Intellectuals and their Strategies for Peace, 1945–1989: Louise Weiss (France), Leo Szilard (USA), E. P. Thompson (England), Danilo Dolci (Italy)*. Chicago: University of Chicago Press, 1993.

Biddle, Tami Davis. *Rhetoric and Reality in Air Warfare: The Evolution of British and American Ideas about Strategic Bombing, 1914–1945*. Princeton: Princeton University Press, 2002.

Bjork, Rebecca S. *The Strategic Defense Initiative: Symbolic Containment of the Nuclear Threat*. Albany: SUNY Press, 1992.

Black, William P., Moses Salomon, and Sigmund Zeisler. *In the Supreme Court of Illinois, North Grand Division. March Term, AD 1887. Brief and Argument for Plaintiffs in Error*. Chicago: Barnhard & Gunthorp, 1887.

Bolt, Neville. *The Violent Image: Insurgent Propaganda and the New Revolutionaries.* New York: Oxford University Press, 2012.

Bonar, James. *Malthus and His Work.* New York: Augustus M. Kelley, 1966.

Bonar, James, C. R. Fay, and J. M. Keynes. "The Commemoration of Thomas Robert Malthus." In *Thomas Robert Malthus: Critical Assessments,* vol. 1, edited by J. C. Wood. Dover: Croom Helm, 1986.

Booth, Wayne C. *The Company We Keep: An Ethics of Fiction.* Berkeley: University of California Press, 1988.

Bown, Stephen R. *A Most Damnable Invention: Dynamite, Nitrates, and the Making of the Modern World.* New York: Thomas Dunne Books, 2005.

Brice, Beatrix. *The Battle Book of Ypres.* New York: St. Martin's, 1988.

"Brigadier General Amos A. Fries." *Chemical Warfare* 3, no. 2 (February 26, 1920): 3–8.

Brodie, Bernard, and Fawn Brodie. *From Crossbow to H-Bomb.* New York: Dell, 1962.

Brophy, Leo P., and George J. B. Fisher. *The Chemical Warfare Service: Organizing for War.* United States Army in World War II. The Technical Services. Washington, DC: Center of Military History, United States Army, 2004.

Brophy, Leo P., Wyndham D. Miles, and Rexmond C. Cochrane. *The Chemical Warfare Service: From Laboratory to Field.* United States Army in World War II. The Technical Services. Washington, DC: Center of Military History, United States Army, 1959.

Brown, Edward M. "Between Cowardice and Insanity: Shell Shock and the Legitimation of the Neuroses in Great Britain." In *Science, Technology and the Military,* vol. 2, edited by Everett Mendelsohn, Merritt Roe Smith, and Peter Weingart, 323–45. Sociology of the Sciences 12.2. Boston: Kluwer, 1988.

Brown, Frederic J. *Chemical Warfare: A Study in Restraints.* Princeton: Princeton University Press, 1968.

Brown, James J., Jr. "The Machine That Therefore I Am." *Philosophy and Rhetoric* 47, no. 4 (2014): 494–514.

Brown, James J., Jr., and Nathaniel A. Rivers. "Composing the Carpenter's Workshop." *O-Zone: A Journal of Object-Oriented Studies* 1 (2013): 27–36. http://o-zone-journal.org/issue.

Brown, Lester R., Gary Gardner, and Brian Halweil. *Beyond Malthus: Nineteen Dimensions of the Population Challenge.* New York: Norton, 1999.

Browne, Stephen Howard. *Angelina Grimké: Rhetoric, Identity, and the Radical Imagination.* East Lansing: Michigan State University Press, 1999.

———. "'This Unparalleled and Inhuman Massacre': The Gothic, the Sacred, and the Meaning of Nat Turner." *Rhetoric and Public Affairs* 3, no. 3 (2000): 309–31.

Brundage, Anthony. *The English Poor Laws, 1700–1930.* New York: Palgrave, 2002.

Bryant, Levi R., and Eileen A. Joy. "Preface: Object/Ecology." *O-Zone: A Journal of Object-Oriented Studies* 1 (2014): i–xiv. http://paas.org.pl/wp-content/uploads/2014/07/bryant-Joy.pdf.

Bryant, Levi R., Nick Srnicek, and Graham Harman. "Towards a Speculative Philosophy." In *The Speculative Turn: Continental Materialism and Realism,* edited by Levi R. Bryant, Nick Srnicek, and Graham Harman, 1–18. Melbourne: re.press, 2011.

Bullough, John D. "Published Works of Theodore Kaczynski." Rensselaer Polytechnic Institute. Accessed January 29, 2018. http://homepages.rpi.edu/~bulloj/tjk/tjk.html.

Bundy, McGeorge. *Danger and Survival: Choices about the Bomb in the First Fifty Years.* New York: Random House, 1988.

Burke, Kenneth. *Attitudes toward History.* 3rd ed. Berkeley: University of California Press, 1984.

———. *A Grammar of Motives*. Berkeley: University of California Press, 1969.

———. *Permanence and Change: An Anatomy of Purpose*. 3rd ed. Berkeley: University of California Press, 1984.

———. *A Rhetoric of Motives*. Berkeley: University of California Press, 1969.

———. "Why Satire, with a Plan for Writing One." *Michigan Quarterly Review* 13, no. 4 (1974): 307–37.

Burrell, George A. "The Research Division, Chemical Warfare Service, U.S.A." *The Journal of Industrial and Engineering Chemistry* 11, no. 2 (1919): 93–104.

Butler, Judith. *Excitable Speech: A Politics of the Performative*. New York: Routledge, 1997.

Campbell, John Angus. "Between the Fragment and the Icon: Prospect for a Rhetorical House of the Middle Way." *Western Journal of Speech Communication* 54, no. 3 (1990): 346–76.

Campbell, Timothy C. *Improper Life: Technology and Biopolitics from Heidegger to Agamben*. Minneapolis: University of Minnesota Press, 2011.

Canaday, John. *The Nuclear Muse: Literature, Physics, and the First Atomic Bombs*. Madison: University of Wisconsin Press, 2000.

Canetti, Elias. *Crowds and Power*. Translated by Carol Stewart. New York: Farrar, Straus and Giroux, 1984.

"Capt. Schaack's Day." *Chicago Tribune*. July 30, 1886.

Ceccarelli, Leah. "Polysemy: Multiple Meanings in Rhetorical Criticism." *Quarterly Journal of Speech* 84, no. 4 (1998): 395–415.

Charles, Daniel. *Master Mind: The Rise and Fall of Fritz Haber, the Nobel Laureate Who Launched the Age of Chemical Warfare*. New York: Ecco, 2005.

Charlesworth, Lorie. *Welfare's Forgotten Past: A Socio-Legal History of the Poor Law*. New York: Routledge, 2010.

Chase, Alston. *A Mind for Murder: The Education of the Unabomber and the Origins of Modern Terrorism*. New York: Norton, 2003.

Chernus, Ira. *Apocalypse Management: Eisenhower and the Discourse of National Security*. Stanford: Stanford University Press, 2008.

Chickering, Roger. "Total War: The Use and Abuse of a Concept." In *Anticipating Total War: The German and American Experiences, 1871–1914*, edited by Manfred F. Boemeke, Roger Chickering, and Stig Förster, 13–28. New York: German Historical Institute and Cambridge University Press, 1999.

Clausewitz, Carl von. *On War*. Edited and translated by Michael Howard and Peter Paret. New York: Knopf, 1993.

Clymer, Jeffory A. *America's Culture of Terrorism: Violence, Capitalism, and the Written Word*. Chapel Hill: University of North Carolina Press, 2003.

Collins, Harry, and Trevor Pinch. *The Golem at Large: What You Should Know about Technology*. New York: Cambridge University Press, 1998.

Conference on the Limitation of Armament Held at Washington. November 12, 1921, to February 6, 1922. Report of the Canadian Delegate Including Treaties and Resolutions. Ottawa: F. A. Acland, 1922.

Corbett, Edward P. J. *Selected Essays of Edward P. J. Corbett*. Edited by Robert J. Conners. Dallas: Southern Methodist University Press, 1989.

Corey, Scott. "On the Unabomber." *Telos* 118 (2000): 157–81.

Cox, J. Robert. "The Die Is Cast: Topical and Ontological Dimensions of the Locus of the Irreparable." *Quarterly Journal of Speech* 68, no. 3 (1982): 227–39.

Crowell, Benedict, and Robert Foster Wilson. *The Armies of Industry, vol. 2. Our Nation's Manufacture of Munitions for a World in Arms, 1917–1918*. New Haven: Yale University Press, 1921.

Cunningham, Andrew, and Ole Peter Grell. *The Four Horsemen of the Apocalypse: Religion, War, Famine and Death in Reformation Europe*. New York: Cambridge University Press, 2000.

Daston, Lorraine. "Introduction: Speechless." In *Things That Talk: Object Lessons from Art and Science*, edited by Lorraine Daston, 9–24. New York: Zone Books, 2004.

David, Henry. *The History of the Haymarket Affair: A Study in the American Social Revolutionary and Labor Movements*. 2nd ed. New York: Russell & Russell, 1958.

"The Declaration of St. Petersburg, 1868." In "Official Documents." *The American Journal of International Law* 1, no. 2. (1907): 95–96.

De Kerckhove, Derrick. "On Nuclear Communication." *Diacritics* 14, no. 2 (1984): 72–81.

Dell, Floyd. "Socialism and Anarchism in Chicago." In *Chicago: Its History and Its Builders*, vol. 2, edited by J. S. Currey, 361–405. Chicago: S. J. Clarke, 1912.

Depew, David. "Introduction to a Special Issue on Rhetorics of Biology in the Age of Biomechanical Reproduction." *POROI* 1, no. 1 (2001): 1–6.

Derrer, Douglas S. *We Are All the Target: A Handbook of Terrorism Avoidance and Hostage Survival*. Annapolis: Naval Institute Press, 1992.

Derrida, Jacques. "No Apocalypse, Not Now (Full Speed Ahead, Seven Missiles, Seven Missives)." *Diacritics* 14, no. 2 (1984): 20–31.

Dery, Mark. *The Pyrotechnic Insanitarium: American Culture on the Brink*. New York: Grove Press, 1999.

Digby, Anne. "Malthus and Reform of the English Poor Law." In *Malthus and His Time*, edited by M. Turner, 157–69. London: Macmillan, 1986.

———. "Malthus and Reform of the Poor Law." In *Malthus Past and Present*, edited by J. Dupâquier, A. Fauve-Chamoux, and E. Grebenik, 97–109. New York: Academic Press, 1983.

Disch, Thomas M. *The Dreams Our Stuff Is Made Of: How Science Fiction Conquered the World*. New York: Free Press, 1998.

Doctor Who. BBC, 2005–.

Doyle, Richard. *On Beyond Living: Rhetorical Transformations of the Life Sciences*. Stanford: Stanford University Press, 1997.

"Drops Balls of Gas on American Lines." *New York Times*, March 21, 1918, 2.

Dupâquier, J., A. Fauve-Chamoux, and E. Grebenik, eds. *Malthus Past and Present*. New York: Academic Press, 1983.

Eden, Lynn, Robert Rosner, Rod Ewing, Sivan Kartha, Edward Kolb, Lawrence M. Krauss, Leon Lederman, et al. "Three Minutes and Counting." *Bulletin of the Atomic Scientists*, January 19, 2015. http://thebulletin.org/three-minutes-and-counting7938.

Ehrlich, Paul R. *The Population Bomb: Population Control or Race to Oblivion?* New York: Ballantine, 1968.

Ellis, John. *The Social History of the Machine Gun*. Baltimore: Johns Hopkins University Press, 1986.

Ellul, Jacques. *Autopsy of Revolution*. Translated by Patricia Wolf. New York: Alfred A. Knopf, 1971.

———. "The Technological Order." Translated by John Wilkinson. In *Philosophy and Technology: Readings in the Philosophical Problems of Technology*, edited by Carl Mitcham and Robert Mackey, 86–105. New York: The Free Press, 1983.

———. *The Technological Society*. Translated by John Wilkinson. New York: Vintage, 1964.

Engels, Friedrich. "Outlines of a Critique of Political Economy." In Karl Marx, *The Economic and Philosophic Manuscripts of 1844*, edited by D. J. Struik, 197–226. New York: International, 1964.

———. "Socialism: Utopian and Scientific." In *The Marx-Engels Reader*, 2nd ed., edited by Robert C. Tucker, 683–717. New York: Norton, 1978.

Erasmus, Desiderius. *De Copia / De Ratione Studii*. In *Collected Works of Erasmus: Literary and Educational Writings* 2, edited by Craig R. Thompson. Toronto: University of Toronto Press, 1978.

Esterer, Arnulf K., and Louise A. Esterer. *Prophet of the Atomic Age: Leo Szilard*. New York: Julian Messner, 1972.

Eureka. NBC Universal Television, 2006–12.

Faith, Thomas I. *Behind the Gas Mask: The US Chemical Warfare Service in War and Peace*. Urbana: University of Illinois Press, 2014.

Farrell, Thomas B. "Knowledge, Consensus, and Rhetorical Theory." *Quarterly Journal of Speech* 62, no. 1 (1976): 1–14.

———. *Norms of Rhetorical Culture*. New Haven: Yale University Press, 1993.

———. "Sizing Things Up: Colloquial Reflection as Practical Wisdom." *Argumentation* 12, no. 1 (1998): 1–14.

Farrell, Thomas B., and G. Thomas Goodnight. "Accidental Rhetoric: The Root Metaphors of Three Mile Island." *Communication Monographs* 48, no. 4 (1981): 271–300.

Feigenbaum, Anna. "Resistant Matters: Tents, Tear Gas, and the 'Other Media' of Occupy." *Communication and Critical/Cultural Studies* 11, no. 1 (2014): 15–24.

Feld, Bernard. "Leo Szilard, Scientist for All Seasons." *Social Research* 51, no. 3 (1984): 675–90.

Fisher, Walter. "Narration as a Human Communication Paradigm: The Case of Public Moral Argument." *Communication Monographs* 51, no. 1 (1984): 1–22.

Foner, Philip S., ed. *The Autobiographies of the Haymarket Martyrs*. New York: Monad Press, 1969.

Foucault, Michel. *The Birth of Biopolitics: Lectures at the Collège de France 1978–1979*. Translated by Graham Burchell. New York: Picador, 2008.

———. *The Birth of the Clinic: An Archaeology of Medical Perception*. Translated by A. M. Sheridan Smith. New York: Vintage, 1994.

———. *The History of Sexuality: An Introduction*. Vol. 1. Translated by Robert Hurley. New York: Vintage, 1990.

———. *"Society Must Be Defended": Lectures at the Collège de France 1975–76*. Translated by David Macey. New York: Picador, 2003.

Fountain, T. Kenny. *Rhetoric in the Flesh: Trained Vision, Technical Expertise, and the Gross Anatomy Lab*. New York: Routledge, 2014.

Francine, Albert P. "Is Chemical Warfare More Inhuman than Gunfire?" *Chemical Warfare* 10, no. 2 (February 15, 1924): 2–6.

Frayn, Michael. *Copenhagen*. New York: Anchor, 2000.

Freedman, Lawrence. *Deterrence*. Malden: Polity, 2004.

Fries, Amos A. "Address by Major General Amos A. Fries, Chief of the Chemical Warfare Service at the Opening Exercise, Army Industrial College, September 1, 1925." National Defense University Library. http://www.ndu.edu/Libraries/Digital-Collections. Accessed January 13, 2011.

———. "By-Products of Chemical Warfare." *The Journal of Industrial and Engineering Chemistry* 20, no. 10 (1928): 1079–84.

———. "Chemical Warfare." *The Journal of Industrial and Engineering Chemistry* 12, no. 5 (1920): 423–29.

———. "Chemical Warfare in Attack." *The Military Engineer* 12, no. 65 (September–October 1920): 454–61.

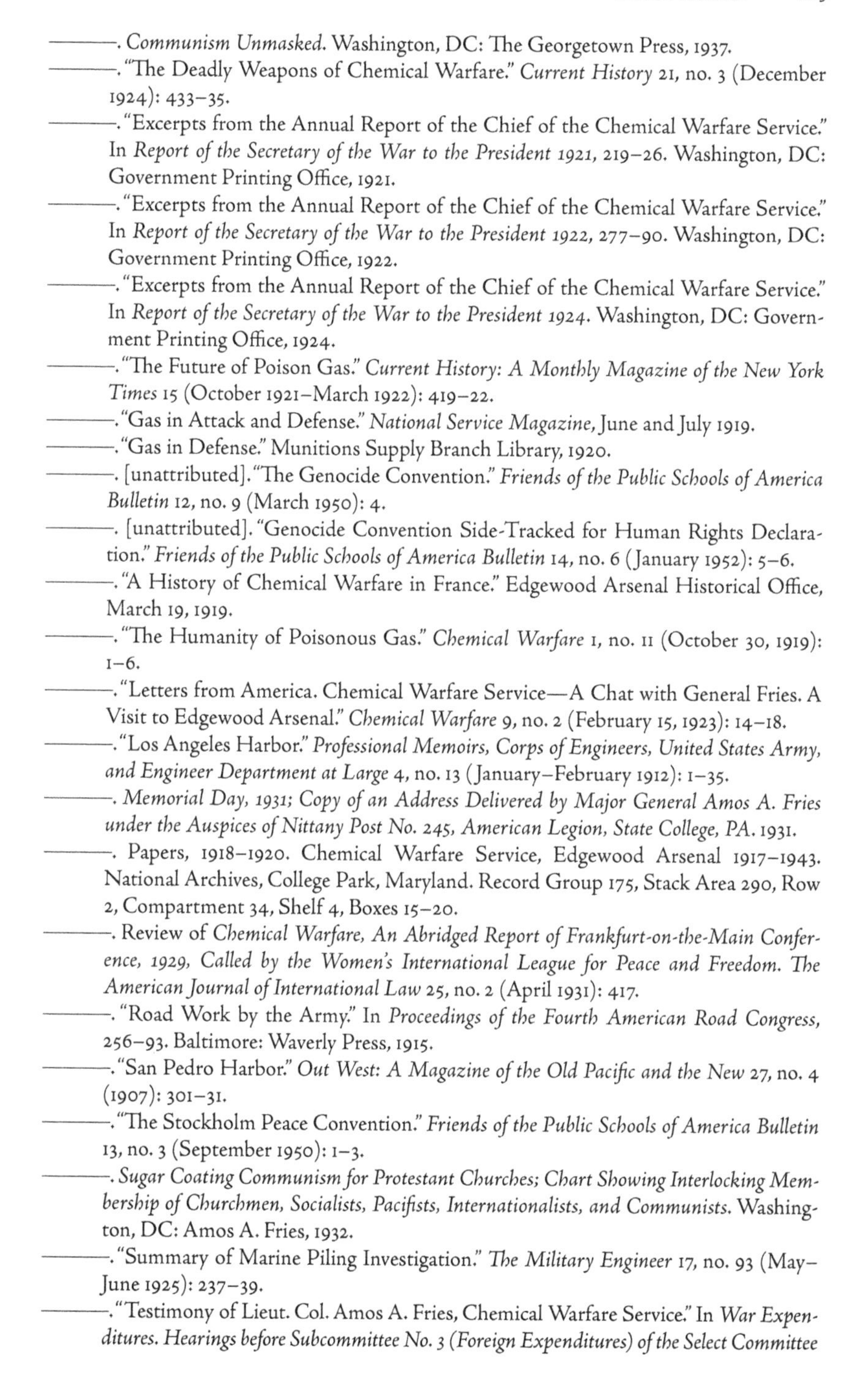

———. *Communism Unmasked*. Washington, DC: The Georgetown Press, 1937.

———. "The Deadly Weapons of Chemical Warfare." *Current History* 21, no. 3 (December 1924): 433–35.

———. "Excerpts from the Annual Report of the Chief of the Chemical Warfare Service." In *Report of the Secretary of the War to the President 1921*, 219–26. Washington, DC: Government Printing Office, 1921.

———. "Excerpts from the Annual Report of the Chief of the Chemical Warfare Service." In *Report of the Secretary of the War to the President 1922*, 277–90. Washington, DC: Government Printing Office, 1922.

———. "Excerpts from the Annual Report of the Chief of the Chemical Warfare Service." In *Report of the Secretary of the War to the President 1924*. Washington, DC: Government Printing Office, 1924.

———. "The Future of Poison Gas." *Current History: A Monthly Magazine of the New York Times* 15 (October 1921–March 1922): 419–22.

———. "Gas in Attack and Defense." *National Service Magazine*, June and July 1919.

———. "Gas in Defense." Munitions Supply Branch Library, 1920.

———. [unattributed]. "The Genocide Convention." *Friends of the Public Schools of America Bulletin* 12, no. 9 (March 1950): 4.

———. [unattributed]. "Genocide Convention Side-Tracked for Human Rights Declaration." *Friends of the Public Schools of America Bulletin* 14, no. 6 (January 1952): 5–6.

———. "A History of Chemical Warfare in France." Edgewood Arsenal Historical Office, March 19, 1919.

———. "The Humanity of Poisonous Gas." *Chemical Warfare* 1, no. 11 (October 30, 1919): 1–6.

———. "Letters from America. Chemical Warfare Service—A Chat with General Fries. A Visit to Edgewood Arsenal." *Chemical Warfare* 9, no. 2 (February 15, 1923): 14–18.

———. "Los Angeles Harbor." *Professional Memoirs, Corps of Engineers, United States Army, and Engineer Department at Large* 4, no. 13 (January–February 1912): 1–35.

———. *Memorial Day, 1931; Copy of an Address Delivered by Major General Amos A. Fries under the Auspices of Nittany Post No. 245, American Legion, State College, PA*. 1931.

———. Papers, 1918–1920. Chemical Warfare Service, Edgewood Arsenal 1917–1943. National Archives, College Park, Maryland. Record Group 175, Stack Area 290, Row 2, Compartment 34, Shelf 4, Boxes 15–20.

———. Review of *Chemical Warfare, An Abridged Report of Frankfurt-on-the-Main Conference, 1929, Called by the Women's International League for Peace and Freedom*. *The American Journal of International Law* 25, no. 2 (April 1931): 417.

———. "Road Work by the Army." In *Proceedings of the Fourth American Road Congress*, 256–93. Baltimore: Waverly Press, 1915.

———. "San Pedro Harbor." *Out West: A Magazine of the Old Pacific and the New* 27, no. 4 (1907): 301–31.

———. "The Stockholm Peace Convention." *Friends of the Public Schools of America Bulletin* 13, no. 3 (September 1950): 1–3.

———. *Sugar Coating Communism for Protestant Churches; Chart Showing Interlocking Membership of Churchmen, Socialists, Pacifists, Internationalists, and Communists*. Washington, DC: Amos A. Fries, 1932.

———. "Summary of Marine Piling Investigation." *The Military Engineer* 17, no. 93 (May–June 1925): 237–39.

———. "Testimony of Lieut. Col. Amos A. Fries, Chemical Warfare Service." In *War Expenditures. Hearings before Subcommittee No. 3 (Foreign Expenditures) of the Select Committee on Expenditures in the War Department. House of Representatives. Sixty-Sixth Congress. Second Session on War Expenditures*, Vol. 2, serial 4, parts 26–50, 1927–56. Washington, DC: Government Printing Office, 1920.

———. "Uses and Dangers of Poisonous Gases; Need for Governmental Regulation." *National Service* 10, no. 2 (August–September 1921): 54–56.

Fries, Amos A., and Clarence J. West. *Chemical Warfare*. New York: McGraw-Hill, 1921.

Fuller, R. Buckminster. *Utopia or Oblivion: The Prospects for Humanity*. New York: Bantam, 1969.

Fuller, Steve. *Philosophy, Rhetoric, and the End of Knowledge: The Coming of Science and Technology Studies*. Madison: University of Wisconsin Press, 1993.

———. "The Strong Program in the Rhetoric of Science." In *Science, Reason, and Rhetoric*, edited by Henry Krips, J. E. McGuire, and Trevor Melia, 95–117. Pittsburgh: University of Pittsburgh Press, 1995.

Future of Life Institute. "Autonomous Weapons: An Open Letter from AI and Robotics Researchers." July 28, 2015. https://futureoflife.org/open-letter-autonomous-weapons.

Gage, Beverly. *The Day Wall Street Exploded: A Story of America in Its First Age of Terror*. New York: Oxford University Press, 2009.

Galison, Peter. "Image of Self." In *Things That Talk: Object Lessons from Art and Science*, edited by Lorraine Daston, 257–94. New York: Zone Books, 2004.

Gandy, David A. *Leo Szilard: Science as a Mode of Being*. New York: University Press of America, 1996.

Gary, Joseph E. "The Chicago Anarchists of 1886: The Crime, the Trial, and the Punishment." *Century Illustrated Magazine*, April, 1893, 803–37.

Gelernter, David. *Drawing Life: Surviving the Unabomber*. New York: Free Press, 1997.

———. *Mirror Worlds: or the Day Software Puts the Universe in a Shoebox . . . How It Will Happen and What It Will Mean*. New York: Oxford University Press, 1991.

Gilbert, G. "Economic Growth and the Poor in Malthus' *Essay on Population*." In *Thomas Robert Malthus: Critical Assessments*, vol. 2, edited by J. C. Wood, 190–202. Dover: Croom Helm, 1986.

Gilbert, Martin. *The First World War: A Complete History*. New York: Henry Holt, 1994.

Gilchrist, H. L. *A Comparative Study of World War Casualties from Gas and Other Weapons*. Washington, DC: Unites States Government Printing Office, 1928.

Gill, Paul, John Horgan, and Jeffrey Lovelace. "Improvised Explosive Device: The Problem of Definition." *Studies in Conflict and Terrorism* 34, no. 9 (2011): 732–48.

Glass, D. V. *Introduction to Malthus*. New York: Wiley & Sons, 1953.

Glebkin, Vladimir. "A Socio-Cultural History of the Machine Metaphor." *Review of Cognitive Linguistics* 11, no. 1 (2013): 145–62.

Godin, Benoit. "Writing Performative History: The New New Atlantis?" *Social Studies of Science* 28, no. 3 (1998): 465–83.

Godwin, William. *Enquiry Concerning Political Justice: And Its Influence on Modern Morals and Happiness*. New York: Penguin, 1985.

———. *Remarks on Mr. Godwin's Inquiry Concerning Population*. London: G. Wilson, 1821.

Goodnight, G. Thomas. "On Questions of Evacuation and Survival in Nuclear Conflict: A Case Study in Public Argument and Rhetorical Criticism." In *Argument in Transition: Proceedings of the Third Summer Conference on Argumentation*, edited by David Zarefsky, Malcolm O. Sillars, and Jack Rhodes, 319–38. Annandale: Speech Communication Association, 1983.

———. "The Personal, Technical, and Public Spheres of Argument: A Speculative Inquiry into the Art of Public Deliberation." *Journal of the American Forensic Association* 18 (Spring 1982): 214–27.

Goodwin, Bridget. *Keen as Mustard: Britain's Horrific Chemical Warfare Experiments in Australia*. St. Lucia: University of Queensland Press, 1998.
Graham, S. Scott. "Agency and the Rhetoric of Medicine: Biomedical Brain Scans and the Ontology of Fibromyalgia." *Technical Communication Quarterly* 18, no. 4 (2009): 289–315.
———. *The Politics of Pain Medicine: A Rhetorical-Ontological Inquiry*. Chicago: University of Chicago Press, 2014.
Green, James. *Death in the Haymarket: A Story of Chicago, the First Labor Movement and the Bombing That Divided Gilded Age America*. New York: Anchor, 2006.
Greene, Ronald Walter. "Another Materialist Rhetoric." *Critical Studies in Mass Communication* 15, no. 1 (1998): 21–41.
———. *Malthusian Worlds: US Leadership and the Governing of the Population Crisis*. Boulder: Westview, 1999.
Gries, Laurie. "Dingrhetoriks." In *Thinking with Bruno Latour in Rhetoric and Composition*, edited by Paul Lynch and Nathaniel Rivers, 294–309. Carbondale: Southern Illinois University Press, 2015.
Griffin, Charles J. G. "'Operation Sunshine': The Rhetoric of a Cold War Technological Spectacle." *Rhetoric and Public Affairs* 16, no. 3 (2013): 521–42.
Grover, David H. *Debaters and Dynamiters: The Story of the Haywood Trial*. Caldwell: Caxton Press, 2006.
Groves, General Leslie M. *Now It Can Be Told: The Story of the Manhattan Project*. New York: Da Capo, 1983.
Gruber, Carol S. "Manhattan Project Maverick: The Case of Leo Szilard." *Prologue: Journal of the National Archives* 15, no. 2 (1983): 72–87.
Haber, Fritz. "The Synthesis of Ammonia from Its Elements." Nobelprize.org, June 2, 1920. http://www.nobelprize.org/nobel_prizes/chemistry/laureates/1918/haber-lecture.pdf.
Haber, L. F. *The Poisonous Cloud: Chemical Warfare in the First World War*. Oxford: Clarendon Press, 1986.
Hacking, Ian. "Biopower and the Avalanche of Printed Numbers." In *Biopower: Foucault and Beyond*, edited by Vernon W. Cisney and Nicolae Morar, 65–81. Chicago: University of Chicago Press, 2016.
———. *The Social Construction of What?* Cambridge, Mass.: Harvard University Press, 1999.
Haldane, J. B. S. *Callinicus: A Defense of Chemical Warfare*. New York: E. P. Dutton, 1925.
Hallenbeck, Sarah. "Toward a Posthuman Perspective: Feminist Rhetorical Methodologies and Everyday Practices." *Advances in the History of Rhetoric* 15, no. 1 (2012): 9–27.
Hammond, James W., Jr. *Poison Gas: The Myths versus Reality*. Westport: Greenwood Press, 1999.
Haraway, Donna. *Simians, Cyborgs, and Women: The Reinvention of Nature*. New York: Routledge, 1991.
Harding, Sandra. *Sciences from Below: Feminisms, Postcolonialities, and Modernities*. Durham: Duke University Press, 2008.
Harman, Graham. "On Vicarious Causation." *Collapse* 2 (2007): 187–211.
The Harvard Committee. *General Education in a Free Society*. Cambridge, Mass.: Harvard University Press, 1945.
Hasian, Marouf, Jr. *Drone Warfare and Lawfare in a Post-Heroic Age*. Tuscaloosa: University of Alabama Press, 2016.
———. "Legal Argumentation in the Godwin-Malthus Debates." *Argumentation and Advocacy* 37, no. 4 (2001): 184–97.
Hawhee, Debra. *Bodily Arts: Rhetoric and Athletics in Ancient Greece*. Austin: University of Texas Press, 2004.

———. *Moving Bodies: Kenneth Burke at the Edges of Language*. Columbia: University of South Carolina Press, 2009.

Hayles, N. Katherine. *How We Became Posthuman: Virtual Bodies in Cybernetics, Literature, and Informatics*. Chicago: University of Chicago Press, 1999.

The Haymarket Affair Digital Collection (HADC). Chicago Historical Society. 2000. http://www.chicagohistory.org/hadc.

Haynes, Williams. *American Chemical Industry, a History*. Vol. 2. New York: D. Van Nostrand, 1945.

Heidegger, Martin. *The Question Concerning Technology and Other Essays*. Translated and edited by William Lovitt. New York: Harper & Row, 1977.

Heiss, Christine. "German Radicals in Industrial America: The Lehr- und Wehr-Verein in Gilded Age Chicago." In *German Workers in Industrial Chicago, 1850–1910: A Comparative Perspective*, edited by Hartmut Keil and John B. Jentz, 206–23. DeKalb: Northern Illinois University Press, 1983.

———. "Popular and Working-Class German Theater in Chicago, 1870–1910." In *German Workers' Culture in the United States 1850–1920*, edited by Hartmut Keil, 181–201. Washington, DC: Smithsonian Institution Press, 1988.

Heller, Charles E. *Chemical Warfare in World War 1: The American Experience, 1917–1918*. Leavenworth: Combat Studies Institute, 1984.

"A Hellish Deed (. . .)" *Chicago Tribune*, May 5, 1886.

Hessel, Frederick A., J. W. Martin, and M. S. Hessel. *Chemistry in Warfare: Its Strategic Importance*. Rev. ed. New York: Hastings House, 1942.

Hikins, James W. "The Rhetoric of 'Unconditional Surrender' and the Decision to Drop the Atomic Bomb." *Quarterly Journal of Speech* 69, no. 4 (1983): 379–400.

Hilgartner, Stephen, Richard C. Bell, and Rory O'Connor. *Nukespeak: The Selling of Nuclear Technology in America*. New York: Penguin, 1983.

Hill, Ian. "'The Human Barnyard' and Kenneth Burke's Philosophy of Technology." *Kenneth Burke Journal* 5, no. 2 (2009). http://www.kbjournal.org/ian_hill.

———. "Memes, Munitions, and Collective Copia: The Durability of the Perpetual Peace Weapons Snowclone." Forthcoming in *Quarterly Journal of Speech* (2018).

———. "Not Quite Bleeding from the Ears: Amplifying Sonic Torture." *Western Journal of Communication* 76, no. 3 (2012): 217–35.

———. "The Rhetorical Transformation of the Masses from Malthus's 'Redundant Population' into Marx's 'Industrial Reserve Army.'" *Advances in the History of Rhetoric* 17, no. 1 (2014): 88–97.

Hobbes, Thomas. *Leviathan*. Mineola: Dover, 2006.

Hoch, Paul K. "The Crystallization of a Strategic Alliance: The American Physics Elite and the Military in the 1940s." In *Science, Technology and the Military*, vol. 1, edited by Everett Mendelsohn, Merritt Roe Smith, and Peter Weingart, 87–116. Sociology of the Sciences 12.1. Boston: Kluwer, 1988.

Hodder, Ian. *Entangled: An Archaeology of the Relationships between Humans and Things*. Malden: Wiley-Blackwell, 2012.

Hoddeson, Lillian, Paul W. Henriksen, Roger A. Meade, and Catherine Westfall. *Critical Assembly: A Technical History of Los Alamos during the Oppenheimer Years, 1943–1945*. New York: Cambridge University Press, 1993.

Hodgson, Geoffrey M. "Malthus, Thomas Robert (1766–1834)." In *Biographical Dictionary of British Economists*, edited by Donald Rutherford. Bristol: Thoemmes Continuum, 2004.

Hogan, J. Michael. *The Nuclear Freeze Campaign: Rhetoric and Foreign Policy in the Telepolitical Age*. East Lansing: Michigan State University Press, 1994.

Hölderlin, Friedrich. *Sämtliche Werke und Briefe*. Edited by Jochen Schmidt. 3 vols. Frankfurt am Main: Deutscher Klassiker Verlag, 1992.

Hollander, Samuel. *The Economics of Thomas Robert Malthus*. Toronto: University of Toronto Press, 1997.

Holley, Irving B. *Ideas and Weapons: Exploitation of the Aerial Weapon by the United States during World War I; A Study in the Relationship of Technological Advance, Military Doctrine, and the Development of Weapons*. Washington, DC: Office of Air Force History, 1983.

Holloway, Rachel L. "The Strategic Defense Initiative and the Technological Sublime: Fear, Science, and the Cold War." In *Critical Reflections on the Cold War: Linking Rhetoric and History*, edited by Martin J. Medhurst and H. W. Brands, 209–32. College Station: Texas A&M University Press, 2000.

Hook, Sidney. *The Fail-Safe Fallacy*. New York: Stein and Day, 1963.

Hughes, Thomas P. "Technological Momentum." In *Does Technology Drive History? The Dilemma of Technological Determinism*, edited by Merritt Roe Smith and Leo Marx, 101–13. Cambridge, Mass.: MIT Press, 1994.

Huzel, James P. *The Popularization of Malthus in Early Nineteenth-Century England: Martineau, Cobbett and the Pauper Press*. Burlington: Ashgate, 2006.

Ihde, Don. *Technology and the Lifeworld: From Garden to Earth*. Bloomington: Indiana University Press, 1990.

"Infernal Machines." *Chicago Tribune*, February 23, 1885.

Irwin, Will. *"The Next War": An Appeal to Common Sense*. New York: E. P. Dutton, 1921.

Isaacson, Edward. *The Malthusian Limit: A Theory of a Possible Static Condition for the Human Race*. London: Methuen, 1912.

Ivie, Robert L. "Metaphor and the Rhetorical Invention of Cold War 'Idealists.'" *Communication Monographs* 54, no. 2 (1987): 165–82.

James, Patricia. *Population Malthus: His Life and Times*. Boston: Routledge & Kegan Paul, 1979.

Jamison, Andrew, and Ron Eyerman. *Seeds of the Sixties*. Berkeley: University of California Press, 1994.

Jasanoff, Sheila. "Ordering Knowledge, Ordering Society." In *States of Knowledge: The Co-Production of Science and Social Order*, edited by Sheila Jasanoff, 13–45. New York: Routledge, 2004.

Johnson, Jeffrey Allan, and Roy MacLeod. "The War the Victors Lost: The Dilemmas of Chemical Disarmament, 1919–1926." In *Frontline and Factory: Comparative Perspectives on the Chemical Industry at War, 1914–1924*, edited by Roy MacLeod and Jeffrey Allan Johnson, 221–45. Dordrecht: Springer, 2006.

Johnson, Jenell. "The Skeleton on the Couch: The Eagleton Affair, Rhetorical Disability, and the Stigma of Mental Illness." *Rhetoric Society Quarterly* 40, no. 5 (2010): 459–78.

Johnson, Sally C. *Psychiatric Competency Report of Dr. Sally C. Johnson, Sept. 11, 1998 in the United States District Court for the Eastern District of California, United States of America, Plaintiff, v. Theodore John Kaczynski, Defendant. CR. NO. S–96–259 GEB.* January 16, 1998.

Jones, Daniel P. "From Military to Civilian Technology: The Introduction of Tear Gas for Civil Riot Control." *Technology and Culture* 19, no. 2 (1978): 151–68.

Jones, Edgar, and Simon Wessely. *Shell Shock to PTSD: Military Psychology from 1900 to the Gulf War*. New York: Psychology Press, 2005.

Jones, Steven E. *Against Technology: From the Luddites to Neo-Luddism*. New York: Routledge, 2006.

Jones, Thai. *More Powerful Than Dynamite: Radicals, Plutocrats, Progressives, and New York's Year of Anarchy*. New York: Bloomsbury, 2012.

Jones, Vincent C. *Manhattan: The Army and the Atomic Bomb*. United States Army in World War II. Special Studies. Washington, DC: Center of Military History, United States Army, 1985.

Joseph, Sister Miriam. *Shakespeare's Use of the Arts of Language*. New York: Columbia University Press, 1947.

Joy, Bill. "Why the Future Doesn't Need Us." *Wired* 8, no. 4, April 2000. https://www.wired.com/2000/04/joy-2.

Jünger, Ernst. "Technology as the Mobilization of the World through the *Gestalt* of the Worker." Translated by J. M. Vincent with R. J. Rundell. In *Philosophy and Technology: Readings in the Philosophical Problems of Technology*, edited by Carl Mitcham and Robert Mackey, 269–89. New York: The Free Press, 1983.

Kaczynski, Theodore J. "Answer to Some Comments Made in *Green Anarchist*." The Anarchist Library. Accessed February 10, 2018. http://theanarchistlibrary.org/library/ted-kaczynski-answer-to-some-comments-made-in-green-anarchist.

———. *The Communiques of Freedom Club*. The Anarchist Library. http://theanarchistlibrary.org/library/ted-kaczynski-the-communiques-of-freedom-club-ted-kaczynski.

———. "Hit Where It Hurts." The Anarchist Library. 2002. http://theanarchistlibrary.org/library/ted-kaczynski-hit-where-it-hurts.

———. *Industrial Society and Its Future*. In *Technological Slavery: The Collected Writings of Theodore J. Kaczynski*, edited by David Skrbina, 36–120. Port Townsend: Feral House, 2010.

———. "Ship of Fools." In *Apocalypse Culture II*, edited by Adam Parfrey, 452–57. Los Angeles: Feral House, 2000.

———. *Technological Slavery: The Collected Writings of Theodore J. Kaczynski*. Edited by David Skrbina. Port Townsend: Feral House, 2010.

———. "Unnamed Essay." 1971.

———. "The Wave of the Future." *The Saturday Review*, June 13, 1970.

———. "When Non-Violence Is Suicide." The Anarchist Library. 2001. http://theanarchistlibrary.org/library/ted-kaczynski-when-non-violence-is-suicide.

———. "Why the Technological System Will Destroy Itself." The Anarchist Library. July 21, 2011. http://theanarchistlibrary.org/library/ted-kaczynski-why-the-technological-system-will-destroy-itself.

Kahn, Herman. "On the Nature and Feasibility of War and Deterrence." In *Armament and Disarmament: The Continuing Dispute*, edited by Walter R. Fisher and Richard Dean Burns, 28–43. Belmont: Wadsworth, 1964.

———. *On Thermonuclear War: Three Lectures and Several Suggestions*. 2nd ed. New York: The Free Press, 1969.

Kant, Immanuel. *Critique of Judgment*. Translated by Werner S. Pluhar. Indianapolis: Hackett, 1987.

———. *Perpetual Peace: A Philosophical Essay*. Translated by M. Campbell Smith. London: Swan Sonnenschein, 1903.

Kaplan, Martin M., Nicholas Kyriakopoulos, S. J. Lundin, J. P. Perry Robinson, and Ralf Trapp. "Summary and Conclusions." In *Verification of Dual-Use Chemicals under the Chemical Weapons Convention: The Case of Thiodiglycol*, edited by S. J. Lundin, 124–36. SIPRI Chemical and Biological Warfare Studies 13. New York: Oxford University Press, 1991.

Kauffman, Charles. "Names and Weapons." *Communication Monographs* 56, no. 3 (1989): 273–85.

Kavka, Gregory S. *Moral Paradoxes of Nuclear Deterrence*. New York: Cambridge University Press, 1987.

Keil, Harmut. "The German Immigrant Working Class of Chicago, 1875–90: Workers, Labor Leaders, and the Labor Movement." In *American Labor and Immigration History, 1877–1920s: Recent European Research*, edited by Dirk Hoerder, 156–76. Urbana: University of Illinois Press, 1983.

———. "Introduction." In *German Workers' Culture in the United States 1850–1920*, edited by Hartmut Keil, xiv–xxii. Washington, DC: Smithsonian Institution Press, 1988.

Keil, Harmut, and Heinz Ickstadt. "Elements of German Working-Class Culture in Chicago, 1880–1890." In *German Workers' Culture in the United States 1850–1920*, edited by Hartmut Keil, 81–108. Washington, DC: Smithsonian Institution Press, 1988.

Keil, Harmut, and John B. Jentz, eds. *German Workers in Chicago: A Documentary History of Working-Class Culture from 1850 to World War I*. Urbana: University of Illinois Press, 1988.

Keith, William, and Kenneth Zagacki. "Rhetoric and Paradox in Scientific Revolutions." *Southern Communication Journal* 57, no. 3 (1992): 165–77.

Keller, Evelyn Fox. *Making Sense of Life: Explaining Biological Development with Models, Metaphors, and Machines*. Cambridge, Mass.: Harvard University Press, 2003.

Kelly, Cynthia C., ed. *The Manhattan Project: The Birth of the Atomic Bomb in the Words of Its Creators, Eyewitnesses, and Historians*. New York: Black Dog & Leventhal, 2007.

Kelly, Kevin. *What Technology Wants*. New York: Viking, 2010.

Kennedy, George A. *Classical Rhetoric and Its Christian and Secular Tradition from Ancient to Modern Times*. 2nd ed. Chapel Hill: University of North Carolina Press, 1999.

Kennedy, Krista. "Textual Machinery: Authorial Agency and Bot-Written Texts in Wikipedia." In *The Responsibilities of Rhetoric*, edited by Michelle Smith and Barbara Warnick, 303–8. Long Grove: Waveland, 2010.

Keränen, Lisa. "Concocting Viral Apocalypse: Catastrophic Risk and the Production of Bio(in)security." *Western Journal of Communication* 75, no. 5 (2011): 451–72.

Kerry, John. "Statement on Syria." US Department of State. August 30, 2013. https://2009-2017.state.gov/secretary/remarks/2013/08/213668.htm.

Kevles, Daniel. "E Pluribus Unabomber: There's a Little of Him in Us All." *The New Yorker*, August 14, 1995.

Keynes, R. "Malthus and Biological Equilibria." In *Malthus Past and Present*, edited by J. Dupâquier, A. Fauve-Chamoux, and E. Grebenik, 359–64. New York: Academic Press, 1983.

Kiesewetter, Renate. "German-American Labor Press: The *Vorbote* and the *Chicagoer Arbeiter-Zeitung*." In *German Workers' Culture in the United States 1850–1920*, edited by Hartmut Keil, 137–55. Washington, DC: Smithsonian Institution Press, 1988.

King, Andrew, and Kenneth Petress. "Universal Public Argument and the Failure of the Nuclear Freeze." *Southern Communication Journal* 55, no. 2 (1990): 162–74.

Kirk, Stuart A., and Herb Kutchins. *The Selling of DSM: The Rhetoric of Science in Psychiatry*. New York: Aldine De Gruyter, 1992.

Kleber, Brooks E., and Dale Birdsell. *The Chemical Warfare Service: Chemicals in Combat*. United States Army in World War II. The Technical Services. Washington, DC: Center of Military History, United States Army, 2003.

Kling, Rob. "Audiences, Narratives, and Human Values in Social Studies of Technology." *Science, Technology, and Human Values* 17, no. 3 (1992): 349–65.

Kropotkin, Peter A. "Anarchism." In *The Encyclopedia Britannica*, 11th ed., vol. 1, 914–19. London: Cambridge University Press, 1911.

———. "Modern Science and Anarchism." In *Anarchism: A Collection of Revolutionary Writings*, edited by R. N. Baldwin, 145–94. Mineola: Dover, 2002.

Kruger, Loren. "Cold Chicago: Uncivil Modernity, Urban Form, and Performance in the Upstart City." *TDR: The Drama Review* 53, no. 3 (2009): 10–36.

Kuhn, Thomas S. "Rhetoric and Liberation." *POROI* 10, no. 2 (2014): Article 4.

———. *The Structure of Scientific Revolutions*. 4th ed. Chicago: University of Chicago Press, 2012.

Kurzweil, Ray. "Promise and Peril." KurzweilAI.net. April 9, 2001. http://www.kurzweilai.net/promise-and-peril.

Kutchins, Herb, and Stuart A. Kirk. *Making Us Crazy: DSM: The Psychiatric Bible and the Creation of Mental Disorders*. New York: Free Press, 1997.

Lange, Jonathan I. "Refusal to Compromise: The Case of Earth First!" *Western Journal of Speech Communication* 54, no. 4 (1990): 473–94.

Lanham, Richard A. *A Handlist of Rhetorical Terms*. 2nd ed. Berkeley: University of California Press, 1991.

Lanouette, William, with Bela Silard. *Genius in the Shadows: A Biography of Leo Szilard, The Man behind the Bomb*. New York: Charles Scribner's Sons, 1992.

Larabee, Ann. *The Wrong Hands: Popular Weapons Manuals and Their Historic Challenge to a Democratic Society*. New York: Oxford University Press, 2015.

Latour, Bruno. *Pandora's Hope: Essays on the Reality of Science Studies*. Cambridge, Mass.: Harvard University Press, 1999.

———. *Science in Action: How to Follow Scientists and Engineers through Society*. Cambridge, Mass.: Harvard University Press, 1988.

———. *We Have Never Been Modern*. Cambridge, Mass.: Harvard University Press, 1993.

Laurence, William L. "Drama of the Atomic Bomb Found Climax in July 16 Test." *New York Times*, September 26, 1945, 1.

Law, John. "Introduction: Monsters, Machines and Sociotechnical Relations." In *A Sociology of Monsters: Essays on Power, Technology and Domination*, edited by John Law, 1–23. New York: Routledge, 1991.

Lazzarato, Maurizio. *Signs and Machines: Capitalism and the Production of Subjectivity*. Translated by Joshua David Jordan. Los Angeles: Semiotext(e), 2014.

Leff, Michael. "Things Made by Words: Reflections on Rhetorical Criticism." *Quarterly Journal of Speech* 78, no. 2 (1992): 223–31.

Lentricchia, Frank, and Jody McAuliffe. *Crimes of Art + Terror*. Chicago: University of Chicago Press, 2003.

Lewis, Michael L. "From Science to Science Fiction: Leo Szilard and Fictional Persuasion." In *The Writing on the Cloud: American Culture Confronts the Atomic Bomb*, edited by Alison M. Scott and Christopher D. Geist, 95–105. New York: University Press of America, 1997.

Locke, John. *An Essay Concerning Human Understanding*. New York: Penguin, 1998.

Luke, Tim. "Re-Reading the Unabomber Manifesto." *Telos* 107 (1996): 81–94.

Lynch, John A., and William J. Kinsella. "The Rhetoric of Technology as a Rhetorical Technology." *POROI* 9, no. 1 (2013): Article 13.

MacDonald, Lyn. *They Called It Passchendaele: The Story of the Third Battle of Ypres and of the Men Who Fought in It*. New York: Atheneum, 1989.

MacKenzie, Donald. *Inventing Accuracy: A Historical Sociology of Nuclear Missile Guidance*. Cambridge, Mass.: MIT Press, 1990.

Malthus, Rev. Thomas R. *Definitions in Political Economy, Preceded by an Inquiry into the Rules Which Ought to Guide Political Economists in the Definition and Use of Their Terms; with Remarks on the Deviation from These Rules in Their Writings*. London: John Murray, 1827.

———. *An Essay on the Principle of Population and a Summary View of the Principle of Population*. New York: Penguin, 1970.

———. *An Essay on the Principle of Population; or a View of Its Past and Present Effects on Human Happiness; with an Inquiry into Our Prospects Respecting the Future Removal or Mitigation of the Evils Which It Occasions*. Edited by Patricia James. 2 vols. New York: Cambridge University Press, 1989.

———. *Occasional Papers of T. R. Malthus: On Ireland, Population, and Political Economy from Contemporary Journals, Written Anonymously and Hitherto Uncollected*. Edited by Bernard Semmel. New York: Burt Franklin, 1963.

———. *The Pamphlets of Thomas Robert Malthus*. New York: Augustus M. Kelley, 1970.

———. *Principles of Political Economy: Considered with a View toward Their Practical Application*, 2nd ed. London: William Pickering, 1836.

Manley, Ron G. "The Problem of Old Chemical Weapons Which Contain 'Mustard Gas or Organoarsenic Compounds: An Overview." In *Arsenic and Old Mustard: Chemical Problems in the Destruction of Old Arsenical and 'Mustard' Munitions*, edited by Josephy F. Bunnett and Marian Mikolajczyk, 1–16. Boston: Kluwer Academic, 1998.

Manning, Van H. *War Gas Investigations: Advance Chapter from Bulletin 178—War Work of the Bureau of Mines*. United States Mines Bureau Bulletin no. 178. Washington, DC: Government Printing Office, 1919.

Marcuse, Herbert. "Repressive Tolerance." In *A Critique of Pure Tolerance*, by Robert Paul Wolff, Barrington Moore Jr., and Herbert Marcuse, 81–123. Boston: Beacon Press, 1969.

Marx, Karl. *Grundrisse: Foundations of the Critique of Political Economy (Rough Draft)*. Translated by Martin Nicolaus. New York: Penguin, 1993.

Marx, Leo. "Technology: The Emergence of a Hazardous Concept." *Technology and Culture* 51, no. 3 (2010): 561–77.

Mauroni, Albert J. *America's Struggle with Chemical-Biological Warfare*. Westport: Praeger, 2000.

May, Matthew S. "Orator-Machine: Autonomist Marxism and William D. 'Big Bill' Haywood's Cooper Union Address." *Philosophy and Rhetoric* 45, no. 4 (2012): 429–51.

Mbembe, Achille. "Necropolitics." Translated by Libby Meintjes. *Public Culture* 15, no. 1 (2003): 11–40.

McConnell, Samuel P. "The Chicago Bomb Case: Personal Recollections of an American Tragedy." *Harper's Monthly Magazine*, May 1934, 730–39.

McCullough, Matthew. "The Higgs Adventure: Five Years In." *CERN Courier* 57, no. 6 (2017): 34–39.

McGee, Michael Calvin. "The 'Ideograph': A Link between Rhetoric and Ideology." *Quarterly Journal of Speech* 66, no. 1 (1980): 1–16.

———. "A Materialist's Conception of Rhetoric." In *Explorations in Rhetoric: Studies in Honor of Douglas Ehninger*, edited by Ray E. Mckerrow, 23–48. Glenview: Scott, Foresman, 1982.

McKeon, Richard. "The Uses of Rhetoric in a Technological Age: Architectonic Productive Arts." In *Rhetoric: Essays in Invention and Discovery*, edited by Mark Backman, 1–24. Woodbridge: Ox Bow Press, 1987.

McLean, George N. *The Rise and Fall of Anarchy in America. From Its Incipient Stage to the First Bomb Thrown in Chicago. A Comprehensive Account of the Great Conspiracy Culminating in the Haymarket Massacre, May 4th, 1886. A Minute Account of the Apprehension, Trial, Conviction and Execution of the Leading Conspirators*. Chicago: R. G. Badoux, 1888.

McNeill, William H. *The Pursuit of Power: Technology, Armed Force, and Society since A.D. 1000*. Chicago: University of Chicago Press, 1982.

Mechling, Elizabeth Walker, and Jay Mechling. "The Campaign for Civil Defense and the Struggle to Naturalize the Bomb." *Western Journal of Speech Communication* 55, no. 2 (1991): 105–33.

Medhurst, Martin J. "Eisenhower's 'Atoms for Peace' Speech: A Case Study in the Strategic Use of Language." *Communication Monographs* 54, no. 2 (1987): 204–20.

———. "Rhetorical Portraiture: John F. Kennedy's March 2, 1962, Speech on the Resumption of Atmospheric Tests." In *Cold War Rhetoric: Strategy, Metaphor, and Ideology*, rev. ed., edited by Martin J. Medhurst, Robert L. Ivie, Philip Wander, and Robert L. Scott, 51–68. East Lansing: Michigan State University Press, 1997.

Meek, Ronald L., ed. *Marx and Engels on the Population Bomb: Selections from the Writings of Marx and Engels Dealing with the Theories of Thomas Robert Malthus*. 2nd ed. Berkeley: Ramparts Press, 1971.

Mello, Michael. "The Non-Trial of the Century: Representations of the Unabomber." *Vermont Law Review* 24, no. 2 (2000): 417–535.

Merton, Robert K. *The Sociology of Science: Theoretical and Empirical Investigations*. Edited by Norman W. Storer. Chicago: University of Chicago Press, 1973.

Michel, Lou, and Dan Herbeck. *American Terrorist: Timothy McVeigh and the Oklahoma City Bombing*. New York: Harper, 2001.

Milbourne, Chelsea Redeker, and Sarah Hallenbeck. "Gender, Material Chronotypes, and the Emergence of the Eighteenth Century Microscope." *Rhetoric Society Quarterly* 43, no. 5 (2013): 401–24.

Miller, Carolyn R. "What Can Automation Tell Us about Agency?" *Rhetoric Society Quarterly* 37, no. 2 (2007): 137–57.

Miller, Richard L. *Under the Cloud: The Decades of Nuclear Testing*. New York: Free Press, 1986.

Mitcham, Carl. *Thinking through Technology: The Path between Engineering and Philosophy*. Chicago: University of Chicago Press, 1994.

Mitchell, Gordon. *Strategic Deception: Rhetoric, Science, and Politics in Missile Defense Advocacy*. East Lansing: Michigan State University Press, 2000.

Molloy, Cathryn. "Recuperative Ethos and Agile Epistemologies: Toward a Vernacular Engagement with Mental Illness Ontologies." *Rhetoric Society Quarterly* 45, no. 5 (2015): 138–63.

Moore, Mark P. "Life, Liberty, and the Handgun: The Function of Synecdoche in the Brady Bill Debate." *Communication Quarterly* 42, no. 4 (1994): 434–47.

Morton, Timothy. "Sublime Objects." *Speculations* 2 (2011): 207–27.

Most, Johann Joseph. *Revolutionäre Kriegswissenschaft; ein Handbüchlein zur Anleitung betreffend Gebrauches und Herstellung von Nitro-Glyzerin, Dynamit, Schiessbaumwolle, Knallquecksilber, Bomben, Brandsätzen, Giften, u.s.w.* New York: International News, 1885.

Mumford, Lewis. *The Myth of the Machine: Technics and Human Development*. New York: Harcourt, Brace & World, 1966.

The National Academies and the Department of Homeland Security. "IED Attack: Improvised Explosive Devices." US Department of Homeland Security. Accessed February 10, 2018. http://www.dhs.gov/xlibrary/assets/prep_ied_fact_sheet.pdf.

Nelson, Bruce C. "*Arbeitspresse und Arbeiterbewegung*: Chicago's Socialist and Anarchist Press, 1870–1900." In *The German-American Radical Press: The Shaping of a Left Political Culture, 1850–1940*, edited by E. Shore, K. Fones-Wolf, and J. P. Danky, 81–107. Urbana: University of Illinois Press, 1992.

———. *Beyond the Martyrs: A Social History of Chicago's Anarchists, 1870–1900*. New Brunswick: Rutgers University Press, 1988.

Nietzsche, Friedrich. *Friedrich Nietzsche on Rhetoric and Language*. Edited by Sander L. Gilman, Carole Blair, and David J. Parent. New York: Oxford University Press, 1989.

Noble, David F. *The Religion of Technology: The Divinity of Man and the Spirit of Invention*. New York: Knopf, 1988.

Nussbaum, Martha. *The Fragility of Goodness: Luck and Ethics in Greek Tragedy and Philosophy*. New York: Cambridge University Press, 2001.

Oakes, Guy. *The Imaginary War: Civil Defense and American Cold War Culture*. New York: Oxford University Press, 1994.

Ogden, C. K., and I. A. Richards. *The Meaning of Meaning: A Study of the Influence of Language upon Thought and the Science of Symbolism*. New York: Harcourt, Brace, 1923.

O'Gorman, Ned. "Eisenhower and the American Sublime." *Quarterly Journal of Speech* 94, no. 1 (2008): 44–72.

———. *Spirits of the Cold War: Contesting Worldviews in the Classical Age of American Security Strategy*. East Lansing: Michigan State University Press, 2012.

Oklahoma Today. *9:02 a.m. April 19, 1995: The Historical Record of the Oklahoma City Bombing*. Oklahoma City: State of Oklahoma, Oklahoma Tourism and Recreation Department, 1996.

Oleson, J. C. "'Evil the Natural Way': The Chimerical Utopias of Henry David Thoreau and Theodore John Kaczynski." *Contemporary Justice Review* 8, no. 2 (2005): 211–28.

O'Mathúna, Dónal P. "Human Dignity in the Nazi Era: Implications for Contemporary Bioethics." *BMC Medical Ethics* 7, no. 2 (2006). doi:10.1186/1472–6939–7–2.

Ong, Walter. *Rhetoric, Romance, and Technology: Studies in the Interaction of Expression and Culture*. Ithaca: Cornell University Press, 1971.

Ott, Brian L., Eric Aoki, and Greg Dickinson. "Ways of (Not) Seeing Guns: Presence and Absence at the Cody Firearms Museum." *Communication and Critical/Cultural Studies*. 8, no. 3 (2011): 215–39.

Packer, Jeremy. "Becoming Bombs: Mobilizing Mobility in the War on Terror." *Cultural Studies* 20, no. 4 (2006): 378–99.

Packer, Jeremy, and Joshua Reeves. "Romancing the Drone: Military Desire and Anthropophobia from SAGE to Swarm." *Canadian Journal of Communication* 38, no. 3 (2013): 309–31.

Parsons, Albert R. "Address of Albert R. Parsons." In *The Chicago Martyrs: The Famous Speeches of the Eight Anarchists in Judge Gary's Court, October 7, 8, 9, 1886, and Reasons for Pardoning Fielden, Neebe and Schwab*, 63–129. San Francisco: Free Society, 1899.

Parsons, Charles L. "The American Chemist in Warfare." *Journal of Industrial Engineering and Chemistry*, no. 10 (1918): 776–80.

Patriquin, Larry. *Agrarian Capitalism and Poor Relief in England, 1500–1860: Rethinking the Origins of the Welfare State*. Basingstoke: Palgrave Macmillan, 2007.

Pechura, Constance M., and David P. Rall, eds. *Veterans at Risk: The Health Effects of Mustard Gas and Lewisite*. Washington, DC: National Academy Press, 1993.

Perelman, Chaim. *The New Rhetoric and the Humanities: Essays on Rhetoric and Its Applications*. Translated by William Kluback. Boston: D. Reidel, 1979.

Perelman, Chaim, and Lucie Olbrechts-Tyteca. *The New Rhetoric: A Treatise on Argumentation*. Translated by John Wilkinson and Purcell Weaver. Notre Dame: Notre Dame University Press, 1969.

Peters, John Durham, and Eric W. Rothenbuhler. "The Role of Rhetoric in the Making of a Science." In *Rhetoric in the Human Sciences*, edited by Herbert W. Simons, 11–27. London: Sage, 1989.

Petersen, William. *Malthus*. Cambridge, Mass.: Harvard University Press, 1979.

———. "Malthus and the Intellectuals." In *Thomas Robert Malthus: Critical Assessments*, vol. 1, edited by J. C. Wood, 366–74. Dover: Croom Helm, 1986.

Petty, Sir William. "Another Essay in Political Arithmetick, Concerning the Growth of the City of London: With the Measures, Periods, Causes, and Consequences Thereof." In *The Economic Writings of Sir William Petty*, vol. 2, edited by Charles Henry Hull, 451–78. Cambridge, Mass.: Cambridge University Press, 1899.

Pickering, Andrew. *The Mangle of Practice: Time, Agency, and Science*. Chicago: University of Chicago Press, 1995.

Plato. *Protagoras*. Translated by K. C. Guthrie. In *The Collected Dialogues of Plato Including the Letters*, edited by Edith Hamilton and Huntington Cairns, 308–52. Princeton: Princeton University Press, 1961.

———. *Socrates' Defense (Apology)*. Translated by Hugh Tredennick. In *The Collected Dialogues of Plato Including the Letters*, edited by Edith Hamilton and Huntington Cairns, 3–26. Princeton: Princeton University Press, 1961.

Poore, Carol. *German-American Socialist Literature, 1865–1900*. Bern: Peter Lang, 1982.

Prelli, Lawrence J. "The Rhetorical Construction of Scientific Ethos." In *Rhetoric in the Human Sciences*, edited by Herbert W. Simons, 48–68. London: Sage, 1989.

Prelli, Lawrence J., Floyd D. Anderson, and Matthew T. Althouse. "Kenneth Burke on Recalcitrance." *Rhetoric Society Quarterly* 41, no. 2 (2011): 97–124.

Prendergast, Catherine. "On the Rhetorics of Mental Disability." In *Embodied Rhetorics: Disability in Language and Culture*, edited by James C. Wilson and Cynthia Lewiecki-Wilson, 45–60. Carbondale: Southern Illinois University Press, 2001.

Preston, Diana. *A Higher Form of Killing: Six Weeks in World War I That Forever Changed the Nature of Warfare*. New York: Bloomsbury, 2015.

Proudhon, Pierre-Joseph. *The Malthusians*. Translated by Benjamin R. Tucker. London: International, 1886.

Pryal, Katie Rose Guest. "The Genre of the Mood Memoir and the *Ethos* of Psychiatric Disability." *Rhetoric Society Quarterly* 40, no. 5 (2010): 479–501.

Pullen, J. M. "Malthus on the Doctrine of Proportions and the Concept of the Optimum." In *Thomas Robert Malthus: Critical Assessments*, vol. 1, edited by J. C. Wood, 419–36. Dover: Croom Helm, 1986.

Quester, George H. *Deterrence before Hiroshima: The Airpower Background of Modern Strategy*. New York: John Wiley & Sons, 1966.

Quintilian. *The Institutio Oratoria of Quintilian*. Translated by H. E. Butler. 4 vols. Cambridge, Mass.: Harvard University Press, 1958.

Rappert, Brian. *Controlling the Weapons of War: Politics, Persuasion, and the Prohibition of Inhumanity*. New York: Routledge, 2006.

Rescher, Nicholas. "Pragmatic Idealism and Metaphysical Realism." In *A Companion to Pragmatism*, edited by John R. Shook and Joseph Margolis, 386–97. Malden: Blackwell, 2006.

———. *A System of Pragmatic Idealism*. 3 vols. Princeton: Princeton University Press, 1992–4.

Restivo, Sal. "4S, the FBI, and Anarchy." *Science, Technology, and Human Values* 26, no. 1 (2001): 87–90.

Rhodes, Richard. *The Making of the Atomic Bomb*. New York: Touchstone, 1988.

Ricardo, David. *The Works and Correspondence of David Ricardo, vol. 8, Letters 1819–June 1821.* Edited by P. Sraffa and M. H. Dobb. New York: Cambridge University Press, 1951.

Richards, I. A. *The Philosophy of Rhetoric.* New York: Oxford University Press, 1936.

Richter, Donald. *Chemical Soldiers: British Gas Warfare in World War I.* Lawrence: University of Kansas Press, 1992.

Rickert, Thomas. *Ambient Rhetoric: The Attunement of Rhetorical Being.* Pittsburgh: University of Pittsburgh Press, 2013.

"The Right Wing Collection of the University of Iowa Libraries 1918–1977: A Guide to the Microfilm Collection." Roosevelt Institute for American Studies, University of Iowa, 1978. https://www.roosevelt.nl.

Rivers, Nathaniel A., and Ryan P. Weber. "Ecological, Pedagogical, Public Rhetoric." *College Composition and Communication* 63, no. 2 (2011): 187–218.

Robinson, J. P. Perry, and Ralf Trapp. "Production and Chemistry of Mustard Gas." In *Verification of Dual-Use Chemicals under the Chemical Weapons Convention: The Case of Thiodiglycol,* edited by S. J. Lundin, 4–21. SIPRI Chemical and Biological Warfare Studies 13. New York: Oxford University Press, 1991.

Rocker, Rudolf. *Nationalism and Culture.* Translated by Ray E. Chase. Los Angeles: Rocker Publications Committee, 1937.

Rohkrämer, Thomas. "How Ernst Jünger Learned to Love Total War." In *The Shadows of Total War: Europe, East Asia, and the United States, 1919–1939,* edited by Roger Chickering and Stig Förster, 179–96. New York: Cambridge University Press, 2003.

Rosemont, Franklin. "A Bomb-Toting, Long-Haired, Wild-Eyed Fiend: The Image of the Anarchist in Popular Culture." In *The Haymarket Scrapbook,* anniversary ed., edited by Franklin Rosemont and David Roediger, 203–12. Chicago: Charles H. Kerr and AK Press, 2011.

Rosemont, Franklin, and David Roediger, eds. *The Haymarket Scrapbook,* anniversary ed. Chicago: Charles H. Kerr and AK Press, 2011.

Rozelle, Lee. *Ecosublime: Environmental Awe and Terror from New World to Oddworld.* Tuscaloosa: University of Alabama Press, 2006.

Russell, Edmund. *War and Nature: Fighting Humans and Insects with Chemicals from World War I to Silent Spring.* New York: Cambridge University Press, 2001.

Scarry, Elaine. *The Body in Pain: The Making and Unmaking of the World.* New York: Oxford University Press, 1985.

Schaack, Michael J. *Anarchy and Anarchists. A History of the Red Terror and the Social Revolution in America and Europe. Communism, Socialism, and Nihilism in Doctrine and in Deed. The Chicago Haymarket Conspiracy, and the Detection and Trial of the Conspirators.* Chicago: F. J. Schulte, 1889.

Schaefermeyer, Mark J. "Dulles and Eisenhower on 'Massive Retaliation.'" In *Eisenhower's War of Words: Rhetoric and Leadership,* edited by Martin J. Medhurst, 27–45. East Lansing: Michigan State University Press, 1994.

Schiappa, Edward. *Defining Reality: Definitions and the Politics of Meaning.* Carbondale: Southern Illinois University Press, 2003.

———. "The Rhetoric of Nukespeak." *Communication Monographs* 56, no. 3 (September 1989): 253–72.

Schüll, Natasha Dow. *Addiction by Design: Machine Gambling in Las Vegas.* Cambridge, Mass.: Princeton University Press, 2014.

Schwarz, L. D. *London in the Age of Industrialization: Entrepreneurs, Labour Force and Living Conditions, 1700–1850.* New York: Cambridge University Press, 1992.

Scott, James Brown, ed. *The Hague Conventions and Declarations of 1899 and 1907 Accompanied by Tables of Signatures, Ratifications and Adhesions of the Various Powers, and Texts of Reservations*. New York: Oxford University Press, 1915.

Scott, Robert L. "Cold War and Rhetoric: Conceptually and Critically." In *Cold War Rhetoric: Strategy, Metaphor, and Ideology*, rev. ed., edited by Martin J. Medhurst, Robert L. Ivie, Philip Wander, and Robert L. Scott, 1–16. East Lansing: Michigan State University Press, 1997.

Self, Lois S. "Rhetoric and *Phronesis*: The Aristotelian Ideal." *Philosophy and Rhetoric* 12, no. 2 (1979): 130–45.

Serbyn, Roman. "Holodomor—The Ukrainian Genocide." *Holodomor Studies* 1, no. 2 (2009): 4–9.

Shaffer, Simon. "A Science Whose Business Is Bursting: Soap Bubbles as Commodities in Classical Physics." In *Things That Talk: Object Lessons from Art and Science*, edited by Lorraine Daston, 147–92. New York: Zone Books, 2004.

Shapin, Steven, and Simon Schaffer. *Leviathan and the Air-Pump: Hobbes, Boyle, and the Experimental Life*. Reprinted ed. Princeton: Princeton University Press, 2011.

Sherwin, Martin J. *A World Destroyed: The Atomic Bomb and the Grand Alliance*. New York: Vintage, 1977.

Shils, Edward. "Leo Szilard: A Memoir." *Encounter* 23, no. 6 (1964): 35–41.

Shrum, Wesley. "We Were the Unabomber." *Science, Technology, and Human Values* 26, no. 1 (2001): 90–101.

Sibert, William. L. *Report of the Chemical Warfare Service 1918*. Washington, DC: Government Printing Office, 1918.

———. *Report of the Chemical Warfare Service 1919*. Washington, DC: Government Printing Office, 1920.

Simons, Herbert W. "Introduction." In *Rhetoric in the Human Sciences*, edited by Herbert W. Simons, 1–9. London: Sage, 1989.

Simonson, Peter. "Reinventing Invention, Again." *Rhetoric Society Quarterly* 44, no. 4 (2014): 299–322.

Skrbina, David. "A Revolutionary for Our Times." In *Technological Slavery: The Collected Writings of Theodore J. Kaczynski*, edited by David Skrbina, 16–34. Port Townsend: Feral House, 2010.

———. "Technological Anarchism: Reconsidering the Unabomber." In *Chromatikon V: Yearbook of Philosophy in Process*, edited by Michel Weber and Ronny Desmet, 189–99. Louvain-la-Neuve: Presses Universitaires de Louvain, 2009.

Sloterdijk, Peter. *Terror from the Air*. Translated by Amy Patton and Steve Corcoran. Los Angeles: Semiotext(e), 2009.

Smith, Adam. *An Inquiry into the Nature and Causes of the Wealth of Nations*. New York: Oxford University Press, 1993.

Smith, Alice K. "The Elusive Dr. Szilard." *Harper's Magazine*, July 1, 1960.

———. *A Peril and a Hope: The Scientists' Movement in America 1945–47*. Rev. ed. Cambridge, Mass.: MIT Press, 1970.

Smith, Carl. *Urban Disorder and the Shape of Belief: The Chicago Fire, the Haymarket Bomb, and the Model Town of Pullman*. Chicago: University of Chicago Press, 1995.

Smith, Kenneth. *The Malthusian Controversy*. London: Routledge & Kegan Paul, 1951.

Solomon, Martha. "Ideology as Rhetorical Constraint: The Anarchist Agitation of 'Red Emma' Goldman." *Quarterly Journal of Speech* 74, no. 2 (1988): 184–200.

Spengler, J. J. "Malthus's Total Population Theory: A Restatement and Reappraisal." In *Thomas Robert Malthus: Critical Assessments*, vol. 2, edited by J. C. Wood, 30–90. Dover: Croom Helm, 1986.

Spies, August. "Address of August Spies." In *The Chicago Martyrs: The Famous Speeches of the Eight Anarchists in Judge Gary's Court, October 7, 8, 9, 1886, and Reasons for Pardoning Fielden, Neebe and Schwab*, 1–16. San Francisco: Free Society, 1899.

———. *August Spies' Autobiography; His Speech in Court, and General Notes*. Edited by Niña Stuart Van Zandt Spies. Chicago: Niña Van Zandt, 1887.

———. *Die Nihilisten*. St. Louis: International Arbeiter Association, 1888.

———. *Reminiszenzen von August Spies. Seine Rede vor Richter Garn, Sozialpolitische Ubhandlungen, Briefe, Notizen, u*. Edited by Albert Currlin. Chicago: Christine Spies, 1888.

Stangeland, Charles Emil. *Pre-Malthusian Doctrines of Population: A Study in the History of Economic Theory*. New York: Columbia University Press, 1904.

Steel, Nigel, and Peter Hart. *Passchendaele: The Sacrificial Ground*. London: Cassell, 2000.

Stormer, Nathan. *Sign of Pathology: US Medical Rhetoric on Abortion, 1800s–1960s*. University Park: Pennsylvania State University Press, 2015.

"Submarines and Gas Condemned by Public." *New York Times*, January 8, 1922, 17.

Sutherland, Ronald G. "Thiodiglycol." In *Verification of Dual-Use Chemicals under the Chemical Weapons Convention: The Case of Thiodiglycol*, edited by S. J. Lundin, 24–43. SIPRI Chemical and Biological Warfare Studies 13. New York: Oxford University Press, 1991.

Szilard, Leo. *The Collected Works of Leo Szilard, Vol. 1: Scientific Papers*. Edited by Bernard T. Feld and Gertrud Szilard, with Kathleen R. Winsor. Cambridge, Mass.: The MIT Press, 1972.

———. *The Collected Works of Leo Szilard, Vol. 2: His Version of the Facts: Selected Recollections and Correspondence*. Edited by Spencer R. Weart and Gertrud Weiss Szilard. Cambridge, Mass.: MIT Press, 1978.

———. *The Collected Works of Leo Szilard, Vol. 3: Toward a Livable World: Leo Szilard and the Crusade for Nuclear Arms Control*. Edited by Helen S. Hawkins, G. Allen Greb, and Gertrud Weiss Szilard. Cambridge, Mass.: MIT Press, 1987.

———. Papers. Special Collections and Archives, University of California, San Diego.

———. *The Voice of the Dolphins and Other Stories*. Expanded ed. Stanford: Stanford University Press, 1992.

Taylor, Bron. "Religion, Violence and Radical Environmentalism: From Earth First! to the Unabomber to the Earth Liberation Front." *Terrorism and Political Violence* 10, no. 4 (1998): 1–42.

Taylor, Bryan C. "The Bodies of August: Photographic Realism and Controversy at the National Air and Space Museum." *Rhetoric and Public Affairs* 1, no. 3 (1998): 331–61.

———. "Home Zero: Images of Home and Field in Nuclear-Cultural Studies." *Western Journal of Communication* 61, no. 2 (1997): 209–34.

Taylor, Michael. "S.F., L.A. Airports Get Bomb Warning/Threat in Letter to Chronicle Appears to Be from Unabomber." *The San Francisco Chronicle*, June 28, 1995. https://www.sfgate.com/news/article/PAGE-ONE-S-F-L-A-Airports-Get-Bomb-Warning-3029895.php.

The Terminator. Directed by James Cameron. Orion Pictures, 1984.

Terminator 2: Judgment Day. Directed by James Cameron. TriStar Pictures, 1991.

Terminator 3: Rise of the Machines. Directed by Jonathan Mostow. Warner Bros. Pictures, 2003.

Tresch, John. *The Romantic Machine: Utopian Science and Technology after Napoleon*. Chicago: University of Chicago Press, 2012.

Turkle, Sherry. *Alone Together: Why We Expect More from Technology and Less from Each Other*. New York: Basic Books, 2012.

United Nations. "Universal Declaration of Human Rights." 1948. http://www.un.org/en/universal-declaration-human-rights/index.html.

United States Bureau of Naval Personnel. *Gas! Know Your Chemical Warfare*. Rev. ed. Washington, DC: United States Bureau of Naval Personnel, 1944.

United States Department of Defense. The Minerva Initiative. Accessed August 9, 2016. http://minerva.defense.gov.

United States Department of Justice. "Procedures for Approving Direct Action against Terrorist Targets Located outside the United States and Areas of Active Hostilities." American Civil Liberties Union. May 22, 2013. https://www.aclu.org/foia-document/presidential-policy-guidance?redirect=node/58033.

United States War Department. *Chemical Warfare Service Field Manual, Volume 1: Tactics and Technique*. Washington, DC: Government Printing Office, 1938.

———. *Technical Manual No. 3–215, Military Chemistry and Chemical Agents, April 22, 1942*. Washington, DC: Government Printing Office, 1943.

Vilensky, Joel A., with Pandy R. Sinish. *Dew of Death: The Story of Lewisite, America's World War I Weapon of Mass Destruction*. Bloomington: Indiana University Press, 2005.

Villard, H. S. "Poison Gas, 1915–1926: German Atrocity—American Necessity." *The Nation*, January 12, 1927, 32.

Virilio, Paul. *City of Panic*. Translated by Julie Rose. New York: Berg, 2005.

———. *The Information Bomb*. Translated by Chris Turner. New York: Verso, 2005.

———. *War and Cinema: The Logistics of Perception*. Translated by Patrick Camiller. New York: Verso, 1989.

Vollmann, William T. "Machines of Loving Grace." *Foreign Policy* 203 (December 2013): 69–71.

Waitt, Alden H. *Gas Warfare: The Chemical Weapon, Its Use, and Protection against It*. New York: Duell, Sloan and Pearce, 1942.

Walsh, Lynda. *Scientists as Prophets: A Rhetorical Genealogy*. New York: Oxford University Press, 2013.

Walters, William. "Drone Strikes, *Dingpolitik* and Beyond: Furthering the Debate on Materiality and Security." *Security Dialogue* 45, no. 2 (2014): 101–18.

Walzer, Arthur E. "Logic and Rhetoric in Malthus's 'Essay on the Principle of Population,' 1798." *Quarterly Journal of Speech* 73, no. 1 (1987): 1–17.

Wander, Philip. "The Rhetoric of American Foreign Policy." In *Cold War Rhetoric: Strategy, Metaphor, and Ideology*, rev. ed., edited by Martin J. Medhurst, Robert L. Ivie, Philip Wander, and Robert L. Scott, 153–83. East Lansing: Michigan State University Press, 1997.

Wang, Jessica. *American Science in an Age of Anxiety: Scientists, Anticommunism, and the Cold War*. Chapel Hill: University of North Carolina Press, 1999.

Warthin, Aldred Scott, and Carl Vernon Weller. *The Medical Aspects of Mustard Gas Poisoning*. St. Louis: C. V. Mosby, 1919.

Weart, Spencer R., and Gertrud Weiss Szilard. Preface to *The Collected Works of Leo Szilard, Vol. 2: His Version of the Facts: Selected Recollections and Correspondence* by Leo Szilard, xvii–xviii. Cambridge, Mass.: MIT Press, 1978.

Weaver, Richard. *The Ethics of Rhetoric*. Davis: Hermagoras Press, 1985.

Wells, H. G. *The World Set Free: A Story of Mankind*. New York: E. P. Dutton, 1914.

West, Clarence J. "The Chemical Warfare Service." In *The New World of Science: Its Development during the War*, edited by Robert M. Yerkes, 148–74. New York: Century, 1920.

———. "The History of Poison Gases." *Science*, May 2, 1919, 412–17.

Whitehead, Ian. "Third Ypres—Casualties and British Medical Services: An Evaluation." In *Passchendaele in Perspective: The Third Battle of Ypres*, edited by Peter H. Liddle, 175–200. London: Leo Cooper, 1997.

Wilson, Thomas. *The Rule of Reason; Conteinying the Arte of Logique*. Edited by Richard S. Sprague. Northridge: San Fernando Valley State College, 1972.

Winch, Donald. "Introduction." In T. R. Malthus, *An Essay on the Principle of Population; or a View of Its Past and Present Effects on Human Happiness; with an Inquiry into Our Prospects Respecting the Future Removal or Mitigation of the Evils Which It Occasions*, edited by Donald Winch, vii–xxiii. New York: Cambridge University Press, 1992.

———. *Malthus*. New York: Oxford University Press, 1987.

Winner, Langdon. *Autonomous Technology: Technics-out-of-Control as a Theme in Political Thought*. Cambridge, Mass.: MIT Press, 1977.

———. "Upon Opening the Black Box and Finding It Empty: Social Constructivism and the Philosophy of Technology." *Science, Technology, and Human Values* 18, no. 3 (1993): 362–78.

Wohlstetter, Albert. "A Delicate Balance of Terror." *Foreign Affairs* 37, no. 2 (1959): 211–34.

Wolff, Robert Paul. "Beyond Tolerance." In *A Critique of Pure Tolerance*, by Robert Paul Wolff, Barrington Moore Jr., and Herbert Marcuse, 3–52. Boston: Beacon Press, 1969.

Woolgar, Steve. "The Turn to Technology in Social Studies of Science." *Science, Technology, and Human Values* 16, no. 1 (1991): 20–50.

Wright, Robert. "The Evolution of Despair." *Time*, June 24, 2001. http://content.time.com/time/magazine/article/0,9171,134603,00.html.

Yellin, Keith. *Battle Exhortation: The Rhetoric of Combat Leadership*. Columbia: University of South Carolina Press, 2008.

Young, Robert M. "Malthus on Man—In Animals No Moral Restraint." In *Malthus, Medicine, and Morality: "Malthusianism" after 1798*, edited by B. Dolan, 73–92. Atlanta: Rodopi, 2000.

Zarefsky, David. *Rhetorical Perspectives on Argumentation: Selected Essays by David Zarefsky*. New York: Springer, 2014.

Zerzan, John. *Running on Emptiness: The Pathology of Civilization*. Los Angeles: Feral House, 2002.

———. *Twilight of the Machines*. Port Townsend: Feral House, 2008.

Index

Lightning Source UK Ltd.
Milton Keynes UK
UKHW011440151119
353513UK00011B/167/P